Energy Efficient Microprocessor Design

Energy Efficient Microprocessor Design

by

Thomas D. Burd
Robert W. Brodersen

with Contributions from
Trevor Pering
Anthony Stratakos

Berkeley Wireless Research Center

University of California,
Berkeley

Distributors for North, Central and South America:
Kluwer Academic Publishers
101 Philip Drive
Assinippi Park
Norwell, Massachusetts 02061 USA
Telephone (781) 871-6600
Fax (781) 681-9045
E-Mail <kluwer@wkap.com>

Distributors for all other countries:
Kluwer Academic Publishers Group
Distribution Centre
Post Office Box 322
3300 AH Dordrecht, THE NETHERLANDS
Telephone 31 78 6392 392
Fax 31 78 6546 474
E-Mail <services@wkap.nl>

 Electronic Services <http://www.wkap.nl>

Library of Congress Cataloging-in-Publication Data

A C.I.P. Catalogue record for this book is available
from the Library of Congress.

Copyright © 2002 by Kluwer Academic Publishers

All rights reserved. No part of this publication may be reproduced, stored in a retrieval system or transmitted in any form or by any means, mechanical, photocopying, recording, or otherwise, without the prior written permission of the publisher, Kluwer Academic Publishers, 101 Philip Drive, Assinippi Park, Norwell, Massachusetts 02061

Printed on acid-free paper.

Printed in the United States of America

To Joyce and Shelagh

Table of Contents

Preface...

Acknowledgements ....

CHAPTER 1 *Introduction* *1*
        1.1 The Need for Energy Efficiency.........1
        1.2 The Performance-Energy Trade-off......3
        1.3 Book Organization 4

CHAPTER 2 *Energy Efficient Design*............................ *7*
        2.1 Processor Usage Model 7
        2.2 CMOS Circuit Models 12
        2.3 Energy Use Metrics 21
        2.4 Energy Efficient Design Observations....30
        2.5 Dynamic Voltage Scaling..............36

CHAPTER 3 *Microprocessor System Architecture*........ *45*
        3.1 System Architecture 46
        3.2 Processor Core 49
        3.3 Cache System 58
        3.4 System Coprocessor 74
        3.5 Summary..........................76

CHAPTER 4 *Circuit Design Methodology* *79*
        4.1 General Energy-Efficient Circuit Design..79

        4.2 Memory Design 92
        4.3 Low-Swing Bus Transceivers 100
        4.4 Design Constraints Over Voltage 109
        4.5 Design Constraints for Varying Voltage . . 116

CHAPTER 5 *Energy Driven Design Flow 125*

        5.1 Overview. 126
        5.2 High-level Energy Estimation 129
        5.3 Clocking Methodology 136
        5.4 Power Distribution Methodology 149
        5.5 Functional Verification. 157
        5.6 Timing Verification 161

CHAPTER 6 *Microprocessor and Memory IC's 169*

        6.1 Microprocessor IC 169
        6.2 Processor Architecture 171
        6.3 Memory IC 211

CHAPTER 7 *DC-DC Voltage Conversion 217*

        7.1 Introduction to Switching Regulators 217
        7.2 PWM Operation 219
        7.3 PFM Operation 227
        7.4 Other Topologies 238
        7.5 Dynamic Voltage Conversion. 241

CHAPTER 8 *DC-DC Converter IC for DVS 251*

        8.1 System and Algorithm Description. 251
        8.2 External Component Selection 257
        8.3 Frequency Detector. 261
        8.4 Current Comparators. 266

　　　　　　　　8.5 Power FETs. 270
　　　　　　　　8.6 Efficiency Simulations 273
　　　　　　　　8.7 Measured Results 275

CHAPTER 9 *DVS System Design and Results 283*

　　　　　　　　9.1 System Architecture 284
　　　　　　　　9.2 Interface IC 285
　　　　　　　　9.3 Prototype Board 290
　　　　　　　　9.4 Software Infrastructure 292
　　　　　　　　9.5 Evaluation 294
　　　　　　　　9.6 Comparisons and other related work 301

CHAPTER 10 *Software and Operating System Support .. 305*

　　　　　　　　10.1 Software Energy Reduction 305
　　　　　　　　10.2 Software Environment 309
　　　　　　　　10.3 System Architecture 312
　　　　　　　　10.4 Benchmarking. 314
　　　　　　　　10.5 DVS Operating System. 322
　　　　　　　　10.6 Voltage Scheduling Algorithms 329
　　　　　　　　10.7 Algorithm Analysis. 338
　　　　　　　　10.8 Comments and Possible Further Directions348

CHAPTER 11 *Conclusions 351*

　　　　　　　　11.1 Energy Efficient Design 351
　　　　　　　　11.2 Current Industry Directions 352
　　　　　　　　11.3 Future Directions 353

　　　　　　　Index 355

Preface

This work began in 1995 as an outgrowth of the InfoPad project which showed us that in order to reduce the energy consumption of a portable multimedia terminal that something had to be done about the consumption of the microprocessor subsystem. The design of the InfoPad attempted to reduce the requirements of this general purpose processor by moving the computation into the network or by the use of highly optimized integrated circuits, but in spite of these efforts it still was a major consumer of energy.

The reasons for this became apparent as we determined that the energy required to perform a function in dedicated hardware could be several orders of magnitude lower than that consumed in the InfoPad microprocessor. We therefore set out on a full fledged attack on all aspects of the microprocessor energy consumption *[1]*. After considerable analysis it became clear that though better circuit design and a streamlined architecture would assist in our goal of energy reduction, that the biggest gains were to be found by operating at reduced voltages. For the busses and I/O this could be accomplished without significant degradation of the processor performance, but this was not a straightforward solution when applied to the core of the processor subsystem (CPU and memory). However, since we could see that a processor in an InfoPad like application (an information appliance access device) only required high performance a relatively small percentage of the time, it became clear that by dynamically varying the voltage to only provide high performance when needed would be a critical component of a complete solution.

This then necessitated further work to support this new degree of flexibility. An efficient DC-DC converter was required that could dynamically and rapidly change the supply voltage. Also software was needed which could predict the required level of performance to provide the illusion of high performance operation, even though most of the time the processor would be running in a more energy efficient, lower voltage mode. Fortunately, Tony Stratakos [2] was pursuing work on efficient CMOS DC-DC convertors and Trevor Pering [3] had been responsible for the software on the InfoPad terminal, so there was expertise which could directly address the new issues stemming from the dynamic voltage operation.

This allowed us to perform a complete system design of the processor starting from the applications through the architecture and finally to the circuits themselves. We felt that a complete implementation was required to demonstrate the concepts and this book contains the complete design process that was developed as well as the integrated circuits that were implemented. We feel that our approach is further validated by recent commercial processors which are using variable voltage operation in the marketplace.

References

[1]   T. Burd, *Energy Efficient Processor System Design*,Ph.D. Thesis, University of California, Berkeley, 2000.

[2]   A. Stratakos, *High-Efficiency, Low-Voltage DC-DC Conversion for Portable Applications*, Ph.D. Thesis, University of California, Berkeley, 1998.

[3]   T. Pering, *Energy-Efficient Operating System Techniques*, Ph.D. Thesis, University of California, Berkeley, 2000.

Acknowledgements

The work would have not been successful without the contributions of other researchers at the University of California, Berkeley. Trevor Pering (Chapter 10) and Tony Stratakos (Chapters 7 and 8) were integral to this work, particularly in demonstrating Dynamic Voltage Scaling on a complete, full-custom, embedded processor system.

Many others helped in the realization of the prototype system. We would like to thank Peggy Laramie and Vandana Prabhu for their help on timing verification, Omid Rowhani for his help functional verification, Patrick Chiang or his help on behavioral modelling, Chris Chang for his help on the standard cell library and clock driver library development, Kevin Camera for his work in the lab verifying the operation of the voltage converter, Hayden So for designing the Xilinx interface on the prototype board, and Sue Mellers for her help designing the test boards.

We would like to thank ARM Ltd. for their support and partnership, and in particular, Simon Segars, for his technical guidance and advice.

We would like to thank Professor Jan Rabaey for his advice and support over the years, and Professor Bora Nikolic for reviewing this material.

This work was supported in part by the Advanced Research Projects Agency and the members of the Berkeley Wireless Research Center.

CHAPTER 1   *Introduction*

1.1 The Need for Energy Efficiency

The explosive proliferation of portable electronic devices has compelled energy-efficient VLSI and system design to provide longer battery run-times, and more powerful products that require ever-increasing computational complexity. In addition, the demand for low-cost and small form-factor devices has kept the available energy supply roughly constant by driving down battery size, despite advances in battery technology which have increased battery energy density. Thus, energy-efficient design must continuously provide more performance per watt.

Since the advent of the integrated circuit (IC), there have been micro-power IC's which have targeted ultra-low-power applications (e.g. watches) with power dissipation requirements in the micro-Watt range *[1.1]*. However, these applications also had correspondingly low performance requirements, and the IC's were not directly applicable to emerging devices (e.g. cell phones, portable computers, etc.) with much higher performance demands.

For the last decade, researchers have made tremendous advancements in energy-efficient VLSI design for these devices, derived, in part, from the earlier micro-power research, and targeted towards the devices' digital signal processing (DSP). Initial work demonstrated how voltage minimization, architectural modification such as parallelism and pipelining, and low-power circuit design could reduce energy consump-

Introduction

tion in low-power custom DSP application-specific IC's (ASIC's) by more than 100x *[1.2]*. Later work demonstrated significant energy-efficiency improvement for a variety of signal-processing applications, including the custom ASIC's in a portable multimedia terminal *[1.3]*, a custom video decoder ASIC *[1.4]*, and programmable DSP IC's *[1.5][1.6][1.7]*.

A common component in these portable devices is a general-purpose processor. Few devices are implemented with a full-custom VLSI solution, as a processor provides two key benefits: the ability to easily implement control functionality which does not map to custom hardware, and more importantly, the ability to upgrade and/or modify functionality, after implementation, due to its programmable nature. Although the processor may perform as little as 1% of the total device computation, advances in energy-efficient custom DSP implementation have made the processor power dissipation a dominant component in portable devices.

Since the advent of the first integrated CMOS microprocessor, the Intel 4004 in 1971, microprocessors were consistently designed with one goal in mind: performance. Processor power and silicon area had been relegated to secondary concern. The widespread emergence of portable devices has created a demand for more energy-efficient processors, but the industry trend has been to fabricate an older processor in a better process technology, operate it at a reduced supply voltage, and market it as a low-power processor. Process and voltage scaling does improve energy-efficiency, but does not provide the improvements possible with a whole-scale processor redesign with minimization of energy consumption considered from the outset.

While some processors have been touted as low-power, and have become quite prevalent in portable devices, they generally achieved this by delivering lower performance. Thus, they are low-power, but not necessarily energy-efficient. The StrongArm processor demonstrated what can be achieved by designing a processor with energy consumption in mind *[1.8]*. It provided a five-fold increase in energy-efficiency, as compared to other contemporary processors, which had otherwise only demonstrated incremental increases.

But even the StrongArm remains 100x-1000x less energy-efficient for basic computation (e.g. arithmetic, logical operations) than a custom ASIC implementation *[1.9]*. This is in large part due to the overhead required by a general-purpose processor: fetching and decoding instructions, multiplexing instructions onto the same underlying hardware, and supporting superscalar and/or pipelined microarchitectures. However, there is still large room for improvement to further improve processor energy-efficiency, as will be demonstrated.

1.2 The Performance-Energy Trade-off

The processor performance and energy consumption is shown in Figure 1.1 for some portable devices currently available. While notebook computer processors can deliver high levels amounts of performance, their high energy consumption requires a relatively large battery to provide just a few hours of run-time. On the other hand, Palm-PCs and PDA's can deliver orders of magnitude longer battery run-time, but it comes at the expense of decreased performance.

In fact, there has been a general performance-energy trade-off, as indicated by the dotted line, which occurs because many existing low-power design techniques sacrifice performance in order to achieve lower power. In DSP applications, parallelism is a common design technique to recover lost performance, but for a general-purpose microprocessor, exploitation of parallelism has been found to be difficult to achieve and to require significant energy consuming overhead.

The most successful approach to increase processor energy efficiency has been to rely on process technology improvements, which increase it by approximately 2x per process generation. In breaking with this philosophy, the designers of the StrongArm processor pushed the trend line up by 5x, and demonstrated that focussing on energy-efficient design could yield as much improvement as two or three process generations could yield.

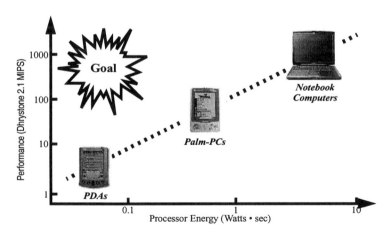

FIGURE 1.1 Processor Performance vs. Energy

Introduction

However, this still falls dramatically short of the goal of a processor that can deliver performance approaching that of a notebook computer, while maintaining PDA-like energy consumption, with energy-efficiencies approaching that which is achievable in low-power, custom ASIC designs.

1.3 Book Organization

The basic organization is to start with a description of the metrics and benchmarks used to design energy-efficient microprocessor systems, which is followed by energy efficient methodologies for the architecture and circuit design, DC-DC conversion, energy efficient software and system integration. Detailed design of the critical circuits and their performance is included throughout.

Chapter 2 presents a usage model for processors found in portable electronic devices in order to qualitatively identify the critical design optimizations for processor performance and energy consumption. Analytical CMOS circuit models are then presented, from which three metrics are derived to quantify energy-efficiency. Four key energy-efficient design principles are presented to demonstrate the application of these metrics. Dynamic voltage scaling (DVS), a technique to dynamically vary a processor's performance and energy consumption, is presented which allows an optimization of these metrics that simultaneously achieves the dual goals of apparent high performance with a high level of energy efficiency.

Chapter 3 presents a top-down energy-efficient architectural design methodology. System-level architectural design issues are discussed, followed by a more in-depth analysis of the processor core and the cache system for the prototype system which is further described in Chapter 6 and Chapter 9.

An energy-efficient circuit design methodology is described in Chapter 4. General circuit design techniques are discussed, including choosing logic styles, transistor sizing, clock-gating, optimizing interconnect, and layout considerations for both standard and datapath cell libraries. Memory design, which has additional constraints placed upon it by DVS, is presented next. Low-swing bus transceivers are then described, which can be used to significantly reduce energy consumption for on-chip busses, and even more significantly, for inter-chip busses.

A methodology for an energy-driven system design flow is described in Chapter 5, which constantly evaluates not only performance, but energy consumption as well, at all levels of the design hierarchy. A majority of the design cycle in modern complex processor designs is spent on validating functionality, a problem which is exacerbated by DVS. The remainder of the chapter describes four parts of the design flow that were developed to aid and speed-up the design of a DVS processor system: clocking methodology, power distribution methodology, functional verification, and timing verification.

In Chapter 6 a detailed description of the design of the core components of a microprocessor system which can implement DVS are described, including both the microprocessor and the memory subsystems. The energy-efficiency trade-off is presented as well as the actual circuits used in the design.

Because of the importance of providing the appropriate value of a supply voltages for maximum energy efficiency, the design of efficient CMOS DC-DC converters are described in Chapter 7. In addition the extra requirements necessary for DVS are described in which the regulator generates a dynamically varying voltage and clock frequency.

In Chapter 8 a complete description of the design of a DC-DC converter for DVS is given which includes the detailed circuits and the optimizations required to achieve high efficiency under dynamic conditions. Additionally, the new metrics required for dynamically varying converters are presented and the results of the prototype circuit are presented.

A complete prototype processor system, consisting of four custom chips, is presented in Chapter 9 which demonstrates the energy-efficiency improvement due to DVS (5-10x), as well as the energy-efficiency improvement due to the previously described energy-efficient design methodology (2-3x).

In order to optimally exploit the advantages of DVS the operating system must be designed to be energy aware. The enhancements required to support DVS are described in Chapter 10, as well as the approach taken for developing the benchmarks which were used in the architecture optimizations and system evaluation.

References

[1.1] E. Vittoz, "Micropower IC's", *Proceedings of the IEEE European Solid-State Circuits Conference*, Sept. 1980, pp. 174-89.

[1.2] A. Chandrakasan, S. Sheng, and R.W. Brodersen, "Low-Power CMOS Digital Design", *IEEE Journal of Solid State Circuits*, Apr. 1992, pp. 473-84.

[1.3] A. Chandrakasan, A. Burstein, and R.W. Brodersen, "A Low Power Chipset for Portable Multimedia Applications", *IEEE Journal of Solid State Circuits*, Vol. 29, No. 12, Dec. 1994, pp. 1415-28.

[1.4] E. Tsern, and T. Meng, "A Low Power Video-rate Pyramid VQ Decoder", *IEEE Journal of Solid-State Circuits*, Vol. 31, No. 11, Nov. 1996, pp. 1789-94.

[1.5] K. Ueda, et. al., "A 16b Low-power-consumption Digital Signal Processor", *Proceedings of the IEEE International Solid-State Circuits Conference*, San Francisco, Feb. 1993, pp. 28-9.

[1.6] T. Shiraishi, et. al., "A 1.8V 36mW DSP for the Half-rate Speech CODEC", *Proceedings of the IEEE Custom Integrated Circuits Conference*, May 1996, pp. 371-4.

[1.7] W. Lee, et. al., "A 1-V Programmable DSP for Wireless Communications", *IEEE Journal of Solid State Circuits*, Vol. 32, No. 11, Nov. 1997, pp. 1766-76.

[1.8] J. Montanaro, et. al., "A 160-MHz 32-b 0.5-W CMOS RISC Microprocessor", *IEEE Journal of Solid State Circuits*, Vol. 31, No. 11, Nov. 1996, pp. 1703-14.

[1.9] H. Zhang, et. al., "A 1-V Heterogeneous Reconfigurable DSP IC for Wireless Baseband Digital Signal Processing", *IEEE Journal of Solid-State Circuits*, Vol. 35, No. 11, Nov. 2000, pp. 1697-1704.

CHAPTER 2 Energy Efficient Design

To effectively optimize the energy efficiency of a processor system, it is critical to first understand the computational demands placed upon it and the usage model of the processor. This information is then coupled with simple CMOS circuit models suitable for deep sub-micron process technologies to define metrics that can be used to derive energy-efficient design principles.

2.1 Processor Usage Model

Understanding a processor's usage pattern is essential to its optimization. Processor utilization can be evaluated in terms of the amount of processing required and the allowable latency for the processing to complete. These two parameters can be merged into the Throughput requirement, T. It is defined as the number of operations that can be performed in a given time:

$$\text{Throughput} \equiv T = \frac{Operations}{Second} \quad \text{(EQ 2.1)}$$

Operations are defined as the basic unit of computation and can be as fine-grained as instructions or more coarse-grained as programs. This leads to measures of throughput of MIPS (instructions/sec) and SPECint95 (programs/sec) [2.1] which compare the throughput on implementations of the same instruction set architecture (ISA), or different ISA's, respectively.

2.1.1 Processor Operation Modes

The desired throughput of various software processes executing on a general purpose processor are shown in Figure 2.1. The example usage pattern shows that the desired throughput varies over time, and the type of computation falls into one of three categories.

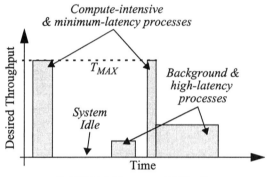

FIGURE 2.1 Processor Utilization.

Compute-intensive and minimum-latency processes desire maximum performance, which is limited by the peak throughput of the processor, T_{MAX}. Any increase in T_{MAX} that the hardware can provide will readily be used by these processes to reduce their latency. Examples of these processes include spreadsheet updates, document spell checks, video decoding, and scientific computation.

Background and high-latency processes require just a fraction of the full throughput of the processor. There is no intrinsic benefit to exceeding the real-time latency requirements of the process since the user will not realize any noticeable improvement. Examples of these processes include video screen updates, data entry, audio/video codecs, and low-bandwidth I/O data transfers.

The third category of computation is system idle, which has zero desired throughput. Ideally, the processor should consume zero power in this mode and therefore be inconsequential. However, in any practical implementation, this is not the case.

Hence, as will be discussed in Section 3.2.6, optimizing this mode of operation requires special attention.

These three modes of operation are found in most single-user processor systems, from personal digital assistants (PDA's), to notebook computers, to powerful desktop machines. This model does not apply to systems implementing a fixed-rate DSP algorithms, which are more efficiently implemented in highly parallel architectures *[2.2][2.3]*. In multi-user mainframe computers, where the processor is constantly in use, this usage model also does not hold true. For these machines, the processor essentially spends the entire time in the compute-intensive mode of operation and it is power dissipation that is often the critical issue. The techniques described in this book to reduce energy will in many cases be applicable to power reduction, particularly those at the circuit and architecture level, but the system level strategies (e.g. DVS) will have less relevance.

2.1.2 What Should be Optimized?

Any increase in processor speed can be readily exploited by compute-intensive and minimum-latency processes. In contrast, the background and high-latency processes do not benefit from any increase in processor speed above and beyond their average desired throughput since the extra throughput cannot be utilized. Thus, peak throughput is the parameter to be maximized since the average throughput is determined by the user and/or operating environment.

The run-time of a portable system is typically constrained by battery life. Simply increasing the battery capacity is not sufficient because the battery has become a significant fraction of the total device volume and weight *[2.4][2.5][2.7]*. Thus, it has become imperative to minimize the load on the battery, while simultaneously increasing the speed of computation to handle ever more demanding tasks. Even for wired desktop machines, the drive towards "green" computers are making energy-efficient design a priority. Therefore, the computation per battery-life/Watt-hour should be maximized, or equivalently, the average energy consumed per operation should be minimized.

Due to the high cost of heat removal, it has also become important to minimize peak energy consumed per operation (i.e. power dissipation), in high-end computing machines and notebook computers. However, the focus of this work is on energy-efficient computing, so the parameter that this work focuses on is average energy consumption rather than peak power dissipation.

Energy Efficient Design

2.1.3 InfoPad: A Case Study in Energy Efficient Computing

The InfoPad is a wireless, multimedia terminal that fits a compact, low-power package in which much of the processing has been moved onto the backbone network *[2.6]*. An RF modem sends/receives data to/from five I/O ports: video output, text/graphics output, pen input, audio input, and audio output. Each I/O port consists of specialized digital IC's, and the associated I/O device (e.g. LCD, speaker, etc.). In addition, there is an embedded processor subsystem used for data flow and network control. InfoPad provides an interesting case study because it contains large amounts of data processing and control processing, which require different optimizations for energy efficiency.

The specialized IC's include a video decompression chip-set which decodes 128×240 pixel frames in real-time, at 30 frames per second. The collection of four chips takes in vector quantized data and outputs analog RGB directly to the LCD and dissipates less than 2mW. Implementing the same decompression in a general purpose processor would require a throughput of around 10 MIPS with hand-optimized code.

The control processing, which has little parallelism to exploit, is much better suited towards a general purpose processor. An embedded processor system for the control functions was therefore designed around the ARM60 processor *[2.8]*, which includes associated SRAM and external glue logic. If this processor were used to deliver a throughput of 10 MIPS it would dissipate1.2W. It is this discrepancy of almost three orders of magnitude in power dissipation that leads to this work's objective of substantially reducing the energy consumption of a general purpose processor system.

2.1.4 The System Perspective

In an embedded processor system such as that found in InfoPad, there are a number of digital IC's external to the processor chip required for a functional system: main memory, clock oscillator, I/O interface(s), and system control logic (e.g., PLD). Integrated solutions have been developed for embedded applications that move the system control logic, the oscillator, and even the I/O interface(s) onto the processor chip leaving only the main memory external such as the SA-1100 processor *[2.9]*.

Figure 2.2 shows a schematic of the InfoPad processor subsystem, which contains the essential system components described above. Interestingly, the processor does not dominate the system's power dissipation; rather, it is the SRAM memory which dissipates half the power. For aggressive energy-efficient design, it is imperative to optimize the entire system and not just a single component; optimizing just the processor in the InfoPad system would yield at most a 10% reduction in overall energy consumption.

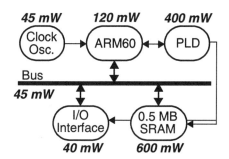

FIGURE 2.2 InfoPad Processor Subsystem.

High-level processor and system simulation is generally used to verify the functionality of an implementation and find potential performance bottlenecks. Unfortunately, such high-level simulation tools do not exist for energy consumption, which forces simulations to extract energy estimates to be delayed until the design has reached the logic design level. At that time, it is very expensive to make significant changes, because system optimizations for energy reduction might require whole-scale redesign or repartitioning.

It is important to understand how design optimizations in one part of a system may have detrimental effects elsewhere. A simple example is the relative effect of a processor's on-chip cache on the external memory system. Because smaller memories have lower energy consumption, the designer may try to minimize the on-chip cache size to minimize the energy consumption of the processor at the expense of a small decrease in throughput (due to increased miss rates of the cache). However, the increased miss rates affect not only the performance, but may increase the system energy consumption as well because high-energy main memory accesses are now made more frequently. So, even though the processor's energy consumption was decreased, the total system's energy consumption has increased.

2.2 CMOS Circuit Models

CMOS has become the predominant process technology for digital circuits. Circuit delays and power dissipation for CMOS circuits can be accurately modeled with simple equations, even for complex processor circuits. These models, along with knowledge about the system architecture, can be used to derive analytical models for energy consumed per operation and peak throughput.

These models will be presented in this section and then used in Section 2.3 to derive metrics that quantify energy efficiency. With these metrics, the circuit and system design can be analytically optimized for maximum energy efficiency.

2.2.1 Power Dissipation

There are four main sources of power dissipation: dynamic switching power due to the charging and discharging circuit capacitances, short-circuit current power due to finite signal rise and fall times, leakage current power from reverse-biased diodes and subthreshold conduction, and static biasing power found in some types of logic styles (i.e. pseudo-NMOS).

Typically, the power dissipation is dominated by the dynamic switching power. However, it is important to understand the other components as they can have a significant contribution to the total power dissipation if proper care is not taken.

Dynamic Switching Power

For every low-to-high output transition in a digital CMOS gate, the capacitance on the output node, C_L, incurs a voltage change ΔV, drawing an energy of $C_L \cdot \Delta V \cdot V_{DD}$ Joules from the supply voltage, V_{DD} *[2.3]*. A high-to-low transition dissipates the energy stored on the capacitor into the NMOS device(s), pulling the output low. The power dissipation is just the product of the energy consumed per transition and the rate at which low-to-high transitions occur, $F_{0 \rightarrow 1}$.

For the simple inverter gate shown in Figure 2.3, ΔV is equal to V_{DD}, so the power drawn from the supply is:

$$\text{Power}_{INVERTER} = C_L \cdot V_{DD}^2 \cdot F_{0 \rightarrow 1} \quad \text{(EQ 2.2)}$$

CMOS Circuit Models

This simple equation holds for more complex gates, and other logic styles as well, given a periodic input. In static logic design, the output only transitions on an input transition, while in dynamic logic, the output is precharged during half of the clock cycle, which may force a transition, with transitions possible in the other half-cycle. In both cases, the power dissipated during switching is proportional to the capacitive load; however, they have different transition frequencies, $F_{0 \to 1}$.

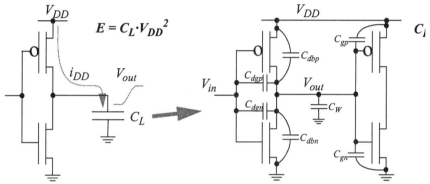

FIGURE 2.3 Dynamic Switching Power Dissipation; Sources of Capacitance.

The basic capacitor elements of C_L, shown in Figure 2.3, consist of the gate capacitance of subsequent inputs attached to the inverter output (C_{gp}, C_{gn}), interconnect capacitance (C_W), and the diffusion capacitance on the drains of the inverter transistors (C_{dbp}, C_{dbn}, C_{dgp}, C_{dgn}) [2.10].

Usually, the value of $F_{0 \to 1}$ is difficult to quantify since it is typically not periodic, and is strongly correlated with the input test vectors. Without doing a transistor-level circuit simulation, the best way to calculate $F_{0 \to 1}$ is to either perform statistical analysis on the circuit[2.11], or use a high-level behavioral model with benchmark software to determine a mean value. Since most digital CMOS circuits are synchronous with a clock frequency, f_{CLK}, an activity factor, $0 < \alpha < 1$, is used to denote the average fraction of clock cycles in which a low-to-high transition occurs, such that $F_{0 \to 1} = \alpha \cdot f_{CLK}$. Therefore, for an integrated circuit with N nodes, the total dynamic switching power is:

$$\text{Power}_{DYNAMIC} = V_{DD} \cdot f_{CLK} \cdot \sum_{i=1}^{N} \alpha_i \cdot C_{Li} \cdot \Delta V_i \qquad \text{(EQ 2.3)}$$

Aside from memory bit-lines and low-swing logic, most nodes swing $\Delta V = V_{DD}$, as was the case for the simple inverter, so that the power equation can be simplified to:

$$\text{Power}_{DYNAMIC} \cong V_{DD}^2 \cdot f_{CLK} \cdot C_{EFF} \qquad \text{(EQ 2.4)}$$

where the effective switched capacitance, C_{EFF}, is commonly expressed as the product of the physical capacitance C_L and the activity weighting factor α, each averaged over the N nodes.

Short-Circuit Current Power

Short-circuit currents occur when the output of a gate is transitioning while the input is still in mid-transition. This generally occurs when the rise/fall time at the input is larger than the output rise/fall time. For the ideal case of a step input, the transistors change state immediately, one turning on, one turning off. There is no conductive path from the supply to ground. For actual circuits, however, the input signal will have a finite rise/fall time. While the conditions $V_{Tn} < V_{in} < V_{DD} - |V_{Tp}|$ and $0 < V_{out} < V_{DD}$ hold for the input/output voltages, there will be a conductive path open because both devices are on.

The longer the input rise/fall time, the longer the short-circuit current will continue to flow, and the average short-circuit current increases. Figure 2.4 plots the increase in energy consumption due to short-circuit current versus the ratio of input rise/fall time (t_{in}) to output rise/fall time (t_{out}) for a static CMOS inverter. The $\Delta E/E_{(tin=0)}$ increases dramatically with increasing input rise/fall time. To minimize the total average short-circuit current power, it is desirable to have equal input and output rise/fall times, since the input rise/fall time of one gate is the output rise/fall time of the previous one.

The average short-circuit current is roughly independent of device size for a fixed load capacitance, since even though the peak magnitude of the current scales with device width, the rise/fall time scales inversely with device width such that the average current is approximately the same. The fraction of power dissipation due to short-circuit current scales with V_{DD}. However, when the supply is lowered to below the sum of the thresholds of the transistors, $V_{DD} < V_{Tn} + |V_{Tp}|$, short-circuit currents will be eliminated because both devices cannot be on at the same time.

For well-designed IC's, the short-circuit power dissipation can be limited to 5-10% of the total dynamic power *[2.12]*. This is achieved by maintaining a bounded ratio on rise/fall times through a transistor-width sizing methodology (Section 4.1.2), so that $\delta_{SC} < .1$ in the following relation:

CMOS Circuit Models

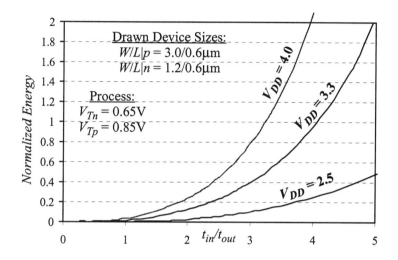

FIGURE 2.4 Short-circuit Energy Consumption vs. Input Rise/fall Time.

$$\text{Power}_{SHORT} = \delta_{SC} \cdot \text{Power}_{DYNAMIC}, \qquad \text{(EQ 2.5)}$$

where δ_{SC} is the ratio of short-circuit to dynamic power dissipation.

Leakage Current Power

There are two types of leakage currents: reverse-bias diode leakage, and sub-threshold leakage through the channel of an "off" device. The magnitude of these currents is set predominantly by the processing technology and total number of transistors.

Diode leakage occurs when one transistor is turned off, and another active transistor changes the drain voltage to create a potential difference with respect to the bulk. For a static CMOS inverter in n-well technology, shown in cross-section in Figure 2.5, with a low input voltage, the output voltage will be high because the PMOS transistor is on. The NMOS transistor will be turned off, but its bulk-to-drain voltage will be equal to the supply voltage, $-V_{DD}$. The resulting diode leakage current will be approximately $I_{LD} = A_D \cdot J_{SD}$, where A_D is the area of the drain diffusion, and J_{SD} is the leakage current density of the diffusion, set by the technology. Since the diode reaches

Energy Efficient Design

maximum reverse-bias current for relatively small reverse-bias potential (< 100mV), the leakage current is roughly independent of supply voltage.

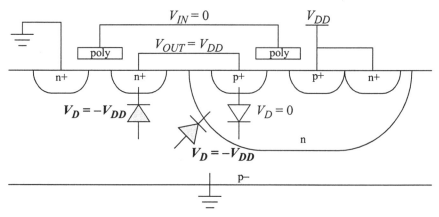

FIGURE 2.5 Reverse-biased Diodes in CMOS Inverter.

In an n-well process, the n-well substrate reverse-biased diode also has leakage current. Since a diode's leakage current is primarily determined by the more lightly doped side of the junction, which is the p– substrate, the leakage current density is similar to that of the NMOS drain-substrate diode [2.13]. Because the well area, A_W, is an order of magnitude larger than the diffusion area, this current will dominate reverse-biased diode leakage in an n-well process. The current is $I_{LW} = A_W \cdot J_{SW}$, where J_{SW} is the leakage current density of the well, also set by the technology.

For the 0.6μm process used here, $J_{SD} \approx 100\text{nA/m}^2$ and $J_{SW} \approx 100\text{nA/m}$ (at 25° C). The leakage current density is temperature sensitive, so J_S can increase dramatically at higher temperatures. Since the well-diode leakage dominates diffusion-diode leakage, the leakage current can be estimated from the size of the die. For a large 200mm² chip, approximately one-half the area is n-well, such that the total diode leakage is on the order of 10pA.

Subthreshold leakage occurs under similar conditions as the diode leakage. In the inverter described above, the NMOS was turned off, but even for $V_{GS} = 0\text{V}$, there is a small current flowing in the channel if V_{DS} is non-zero. The I_D vs. V_{GS} characteristic,

CMOS Circuit Models

shown in Figure 2.6, shows the exponential relation which is characteristic of this subthreshold region ($V_{GS} < |V_T|$).

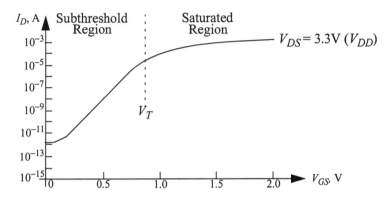

FIGURE 2.6 I_D vs. V_{GS} for MOSFET in Subthreshold Region.

The magnitude of the subthreshold current is both a function of process, device sizing (W/L), and supply voltage [2.14]. The process parameter that predominantly affects the current value is V_T. For a V_T of 0.8V, the current magnitude for a single device is on the order of 1pA. Approximately one out of every two transistors has the necessary bias conditions for subthreshold leakage, so that for a 2 million transistor chip, the total subthreshold current would be on the order of 1µA, which dwarfs the reverse-diode leakage current.

The combination of diode-leakage current and subthreshold current for a 2 million transistor chip is therefore approximately 1µA, which at a supply voltage of 3.3V, is below 10µW. This is insignificant to the dynamic switching power while the processor is operating. This power is only important in setting the lower threshold of achievable power dissipation while the processor is idling. Hence, this power component will be ignored except when discussing idle energy consumption.

However, as process technology continues to advance, the maximum operating voltage decreases, and reductions in V_T are required to maintain a reasonable gate-drive voltage. This is particularly true in processes targeted towards high-performance IC's. For example, a 0.35µm process with $V_T = 0.35V$ has a leakage current on the order of 10nA per device [2.15][2.16]. This yields 10mA of leakage current for the same 2 million transistor chip, and may become a significant fraction of the power dissipation in a very low-power chip. There are a number of design techniques to reduce this leakage current, such that it is once again only critical when considering idle energy consumption, such as selectively increasing channel lengths and dual V_T devices *[2.17]*, and dynamically varying V_T *[2.18]*.

Static Biasing Power

While static bias currents are usually avoided in CMOS circuits, occasionally, they may prove to be beneficial. A typical application is for a large complex gate that cannot be implemented with dynamic logic due to asynchronous timing constraints.

Figure 2.7 contains an example gate; it is a wide AND-OR-Invert gate with asynchronous inputs. To implement this in full static CMOS would require several times the area to implement the stacked PMOS transistors. The extra PMOS transistors would also increase the capacitance on the input nodes, loading down the previous gates. However, by synchronizing the inputs through architectural design, which can usually be accomplished, and then implementing the complex gates with dynamic logic, this power component can be made negligible.

Combined Power Model

With the assumption that no static biasing is present, the total power dissipation is just the summation of the remaining three individual components:

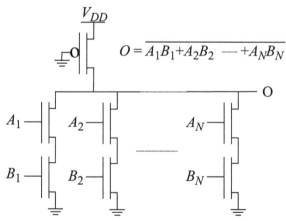

FIGURE 2.7 Implementing Complex Logic with Static Biasing (pseudo-NMOS)

CMOS Circuit Models

$$\text{Power} = \text{Power}_{DYNAMIC} + \text{Power}_{SHORT} + \text{Power}_{LEAKAGE} \quad \text{(EQ 2.6)}$$

where the $\text{Power}_{LEAKAGE}$ component is on the order of 10-100μW.

$$\text{Power} = (1 + \delta_{SC}) \cdot V_{DD}^2 \cdot f_{CLK} \cdot C_{EFF} + \text{Power}_{LEAKAGE} \approx V_{DD}^2 \cdot f_{CLK} \cdot C_{EFF} \quad \text{(EQ 2.7)}$$

To simplify the following analyses, the assumptions that $(1+\delta_{SC}) \approx 1$, and that $\text{Power}_{LEAKAGE}$ can be ignored except during processor idle will be made, so that the total chip power dissipation is approximately equal to just the dynamic switching power component.

2.2.2 Energy per Operation

A common measure of energy consumption is the power-delay product (PDP) [2.2]. This delay is often defined as the critical path delay, so PDP is equivalent to the energy consumed per clock cycle (Power / f_{CLK}). However, the measure of interest is the energy consumed per operation which can be derived by dividing the PDP by the number of operations per clock cycle. The energy consumed per operation can now be expressed as a function of effective switched capacitance, supply voltage, and operations per clock cycle:

$$\frac{\text{Energy}}{\text{Operation}} \cong \frac{V_{DD} \cdot C_{EFF}}{\text{Operations / Clock Cycle}} \quad \text{(EQ 2.8)}$$

2.2.3 Circuit Delay

To fully utilize its hardware, a digital circuit should be operated at the maximum possible frequency. This maximum frequency is just the inverse of the delay of the processor's critical path which is proportional to the delay of a single CMOS gate

The delay for a CMOS gate, which is defined as the time required for the output to transition to 50% of the voltage swing, V_{DD}, can be approximated as [2.10]:

$$\text{Delay} \cong \frac{C_L}{I_{AVE}} \cdot \Delta V_{0\% \to 50\%} = \frac{C_L}{I_{AVE}} \cdot \frac{V_{DD}}{2} \quad \text{(EQ 2.9)}$$

Energy Efficient Design

I_{AVE} is the average device current during the transition. For sub-micron MOS devices in velocity saturation, the device current, I_D, is [2.19]:

$$I_D = v_{SAT} \cdot C_{OX} \cdot W \cdot (V_{DD} - V_T - V_{Dsat}) \qquad V_{DS} \geq V_{Dsat} \qquad \text{(EQ 2.10)}$$

with the assumption of a fast input signal transition so that the device's gate-source voltage is V_{DD}. The term v_{SAT} is the maximum carrier velocity, C_{OX} is the gate capacitance, W is the device width, and V_T is the device threshold. V_{Dsat} is the value above which the current is independent of the drain-source voltage, V_{DS}, which in velocity saturation is given by:

$$V_{Dsat} = \left(\frac{E_C \cdot L_e}{V_{DD} - V_T + E_C \cdot L_e}\right)(V_{DD} - V_T) \qquad \text{(EQ 2.11)}$$

where L_e is the effective electrical channel length, while E_C is the longitudinal electrical field at which the carriers are considered at v_{SAT}, and is a fundamental constant of silicon. E_C is approximately 1.5×10^6 V/m [2.20], so with the reasonable approximation that $V_T \sim E_C L_e$ (e.g. if $L_e = 0.5\mu m$, $V_T = 0.75V$):

$$V_{Dsat} \approx \left(\frac{V_T}{V_{DD}}\right)(V_{DD} - V_T) \qquad \text{(EQ 2.12)}$$

The device remains in saturation during the output transition (defined as a 50% change in output voltage), since V_{Dsat} is well below $\frac{1}{2}V_{DD}$ for the entire range of V_{DD}, and the current is approximately constant such that $I_{AVE} = I_D$. Combining equations (EQ 2.12), (EQ 2.10), and (EQ 2.9) yields:

$$\text{Delay} \cong \frac{C_L \cdot V_{DD}^2}{k_v \cdot W \cdot (V_{DD} - V_T)^2} \qquad k_v = 2 \cdot v_{SAT} \cdot C_{OX} \qquad \text{(EQ 2.13)}$$

where k_V contains the technology dependence. For our 0.6μm process ($L_e = 0.4$, $V_T = 0.75V$), this approximation gives less than 10% error in comparison to a SPICE-simulated delay using a BSIM3 device model [2.21].

2.2.4 Throughput

Throughput was previously defined in Equation 2.1 as the number of operations that can be performed in a given time. When clock rate is set to be the inverse of the critical path delay, the throughput is equal to the amount of computational concurrency (i.e. operations completed per clock cycle) divided by this delay:

$$T = \frac{Operations}{Second} = \frac{Operations\ per\ Clock\ Cycle}{Critical\ Path\ Delay} \quad \text{(EQ 2.14)}$$

The critical path delay can be related back to the previous delay model by summing up the delay over all M gates in the critical path:

$$Critical\ Path\ Delay \cong \frac{V_{DD}^2}{k_v \cdot (V_{DD} - V_T)^2} \cdot \sum_{i=1}^{M} \frac{C_{Li}}{W_i} \quad \text{(EQ 2.15)}$$

Making the approximation that all gate delays are equal, Equation 2.15 can be simplified if N_{gates} is used to indicate the length of the critical path (i.e. number of gates), and average values for C_L and W are used. Throughput can now be expressed as a function of a technology parameter, supply voltage, critical path length, and operations per clock cycle:

$$T \cong \frac{k_v \cdot W \cdot (V_{DD} - V_T)^2}{N_{gates} \cdot C_L \cdot V_{DD}^2} \cdot \frac{Operations}{Clock\ Cycle} \quad \text{(EQ 2.16)}$$

Typical units for operations per clock cycle are MIPS/Mhz, and SPECint95/MHz when operations are respectively defined as instructions and benchmark programs.

2.3 Energy Use Metrics

While the energy consumed per operation should always be minimized, no single metric quantifies the efficiency of energy use for all digital systems, since the metric must be dependent on the system's throughput constraints. Three main modes of computation can be defined which include fixed throughput, maximum throughput, and burst throughput. Each of these modes can be characterized by a simple metric for energy use, as described in the following three sections. While single-user systems typically operate in the burst throughput mode, the other two modes are equally important since they are degenerate forms of the burst throughput mode in which the system may be required to operate.

2.3.1 Fixed Throughput Mode

Many real-time systems require a fixed number of operations per second. Any excess throughput cannot be utilized, and therefore needlessly consumes energy. Systems with this characteristic will be defined as operating in the fixed throughput mode of computation, and they are typically found in digital signal processing applications in which the required throughput is set by a fixed-rate incoming or outgoing real-time signal (e.g., speech, audio, video).

$$Energy\ Use\ Metric\big|_{FIX} = \frac{Power}{Throughput} = \frac{Energy}{Operation} \quad \text{(EQ 2.17)}$$

Previous work has shown that the optimization of the metric in Equation 2.17 is valid for the fixed throughput mode of computation [2.2]. A lower value implies a more energy-efficient solution. If a design can be made twice as energy efficient (i.e. reduce the energy/operation by a factor of two), then its sustainable battery life has been doubled, and since the throughput is constant, its power dissipation has been halved. For the fixed-throughput mode, minimizing the power dissipation is equivalent to minimizing the energy/operation.

Using the energy model in Section 2.2.2, the metric can be expressed as:

$$Energy\ Use\ Metric\big|_{FIX} = \frac{V_{DD}^2 \cdot C_{EFF}}{Operations\ /\ Clock\ Cycle} \quad \text{(EQ 2.18)}$$

The primary way to improve energy efficiency is to reduce supply voltage while maintaining the throughput constraint, which yields a quadratic improvement in energy efficiency. Additionally, reducing the effective switched capacitance will also improve efficiency. Optimizing the energy efficiency of this mode of computation has been the focus of much previous work, which has yielded a variety of low-power design techniques that provide significant efficiency improvements [2.3].

2.3.2 Maximum Throughput Mode

In most multi-user systems, primarily networked workstations and mainframes, the processor is continuously running. The faster the processor can perform computation, the better, yielding the defining characteristic of the maximum throughput mode of computation. Thus, this mode's metric of energy use must balance the need for low energy/operation and high throughput, which is accomplished through the optimization of the Energy-to-Throughput Ratio, or *ETR*:

$$Energy\ Use\ Metric|_{MAX} = ETR = \frac{E_{MAX}}{T_{MAX}} = \frac{Power}{Throughput^2} \qquad \text{(EQ 2.19)}$$

where E_{MAX} is the energy/operation, or equivalently power for the maximum throughput operation, and T_{MAX} is the maximum throughput achievable.

A lower *ETR* indicates lower energy/operation for equal throughput, or equivalently, indicates greater throughput for the same amount of energy/operation, satisfying the need to equally optimize throughput and energy/operation. Thus, a lower *ETR* represents a more energy-efficient solution and represents the amount of energy use normalized to the throughput.

The Energy-Delay Product (EDP) is a similar metric*[2.22]*, but does not include the effects of architectural parallelism when the delay is taken to be the critical path delay. For example, two processors may consume the same energy/operation and operate at the same clock frequency, but one processor can be designed to complete two operations per cycle, while a second processor can only complete one operation. Since the EDP for the two processors is the same, this incorrectly indicates that they might have equivalent energy use characteristics. However, since the first processor has twice the throughput of the second, its energy to throughput ratio, ETR, will be one-half of the second processor correctly indicating that the processor which can complete two operations per clock cycle is actually twice as efficient in energy use when throughput is held constant.

Throughput and energy/operation can be scaled with supply voltage, as shown in Figure 2.8 (the data for Figures Figure 2.8-Figure 2.10 is derived from Equation 2.8 and Equation 2.16, which models sub-micron CMOS processes); but, unfortunately, they do not scale proportionally. So while throughput and energy/operation can be

Energy Efficient Design

varied by well over an order of magnitude to cover a wide dynamic range of operating points, the *ETR* is not constant for different values of supply voltage.

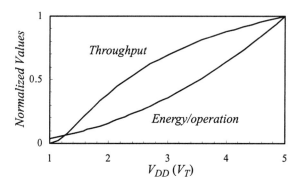

FIGURE 2.8 Energy/operation, Throughput vs. V_{DD}.

As shown in Figure 2.9, V_{DD} can be adjusted by a factor of 2.5 (1.4-3.5V_T) and the *ETR* only varies within 50% of the minimum at $2V_T$. However, outside this range, the *ETR* rapidly increases. Clearly, for supply voltages greater than $4V_T$ there is a rapid increase in the energy required to sustain a given throughput, as well as for supply voltages that approach the device threshold voltage. But, since both throughput and energy/operation are monotonically increasing function of supply voltage, varying V_{DD} allows throughput to be traded off for lower energy/operation, and vice-versa.

To compare designs over a larger operating range for the maximum throughput mode, a better metric is a plot of the energy/operation versus throughput. To make this plot, the supply voltage is varied from the minimum operating voltage (near V_T in many digital CMOS designs) to the maximum voltage (2.5-5V, depending on the technology), while energy/operation and throughput are measured. The energy/operation can then be plotted as a function of throughput, and the architecture is completely characterized over all possible throughput values.

Using a single value for *ETR* is equivalent to making a linear approximation to the actual energy/operation versus throughput curve. Figure 2.10 demonstrates the error incurred in using a constant *ETR* metric, which is calculated at a nominal supply voltage of 3.3V for this example. For architectures with similar throughput, a single *ETR* value is a reasonable metric for normalized energy use; however, for designs opti-

Energy Use Metrics

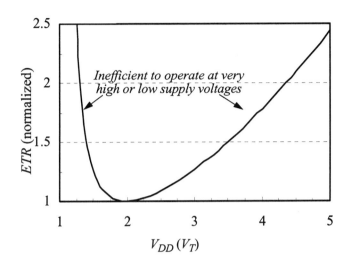

FIGURE 2.9 ETR as a function of V_{DD}.

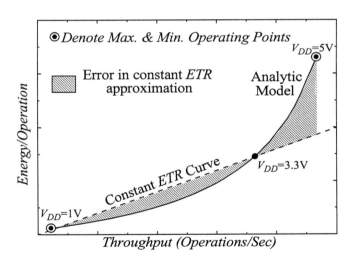

FIGURE 2.10 Energy use vs. Throughput metric.

mized for vastly different values of throughput, a plot may be more useful, as Section 2.4.1 will demonstrate.

Using the throughput and energy models from Section 2.2, the ETR is:

$$ETR = \frac{C_L \cdot N_{gates} \cdot C_{EFF} \cdot V_{DD}^4}{k_v \cdot W \cdot (V_{DD} - V_T)^2} \qquad \text{(EQ 2.20)}$$

This equation must be carefully used in providing insight into energy-efficient design, since the variables have interdependencies. For example, if the device width, W, is increased to reduce ETR, C_L and C_{EFF} will also increase, effectively increasing ETR when the gate capacitance begins to dominate the load capacitance. Similarly, if N_{gates} is reduced, this may come at the cost of increased C_L and/or C_{EFF}. Hence, individual parameters cannot be optimized in isolation, and their inter-dependencies must be taken into account by fully evaluating the ETR. However, this formulation provides a framework for evaluating various alternatives when designing for energy efficiency with the throughput implicitly taken into account.

2.3.3 Burst Throughput Mode

Most single-user systems (e.g., stand-alone desktop computers, notebook computers, PDA's, etc.) spend only a fraction of the time performing useful computation. The rest of the time is spent idling between processes. However, when bursts of computation are demanded, the faster the throughput (or equivalently, response time), the better. This characterizes the burst throughput mode of computation in which most portable devices operate. The metric for energy efficiency used for this mode must balance the desire to minimize energy consumption, while both idling and computing, and to maximize peak throughput when computing.

Ideally, the processor's clock should track the periods of computation in this mode so that when an idle period is entered, the clock is immediately shut off. Then a good metric of energy use is just the energy throughput ratio, ETR, as the energy consumed while idling has been eliminated. However, this is not realistic in practice, since many processors do not having an energy saving mode and those that do so generally support only simple clock reduction/deactivation modes.

The hypothetical example depicted in Figure 2.11 contains a clock reduction (sleep) mode in which major sections of the processor are shut down. The shaded area indicates the processor's idle cycles in which energy is needlessly consumed, and whose

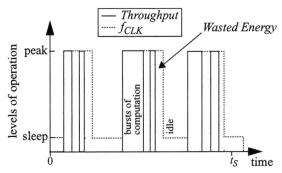

FIGURE 2.11 Wasted energy due to idle cycles.

magnitude is dependent upon whether the processor is operating in the "low-power" mode. The energy/operation while actively computing, E_{MAX}, and the amortized energy/operation while idling, E_{IDLE}, is:

$$E_{MAX} = \frac{\text{Total Energy Consumed Computing}}{\text{Total Operations}} \quad \text{(EQ 2.21)}$$

$$E_{IDLE} = \frac{\text{Total Energy Consumed Idling}}{\text{Total Operations}} \quad \text{(EQ 2.22)}$$

Total energy and total operations are calculated over the sample time period, t_S. T_{MAX} is the peak throughput during the bursts of computation (as defined in Section 2.3.2), and T_{AVE} is the time-averaged throughput (total operations / t_S). The time period, $t_{S,}$ is chosen to be long enough that a good estimate of the "average" throughput requirements, $T_{AVE,}$ of the user or target system environment can be made then a good metric of energy efficiency for the burst throughput mode is:

$$BETR = \frac{E_{MAX} + E_{IDLE}}{T_{MAX}} \quad \text{(EQ 2.23)}$$

This metric will be called the Burst-mode *ETR* (*BETR*); it is similar to *ETR*, but also accounts for energy consumed while idling. Again a lower *BETR* represents a more energy-efficient solution.

Multiplying Equation 2.21 by the actual time computing, (T_{AVE} / T_{MAX}), shows that E_{MAX} is the ratio of compute power dissipation to peak throughput T_{MAX}, as previously defined in Section 2.3.2. Thus, E_{MAX} is only a function of the hardware and can be measured by operating the processor at full utilization.

E_{IDLE}, however, is a function of t_S and T_{AVE}. The power consumed idling must be measured while the processor is operating under typical conditions, and T_{AVE} must be known to then calculate E_{IDLE}. E_{IDLE} can therefore be expressed as a function of the idle power dissipation, the time spent idling during a sample period, as well as t_S and T_{AVE} so that Equation 2.22 can be rewritten as:

$$E_{IDLE} = \frac{[Idle\ Power\ Dissipation][Time\ Idling]}{[Average\ Throughput][Sample\ Time]} \quad \text{(EQ 2.24)}$$

However, expressing E_{IDLE} as a function of E_{MAX} better illustrates the conditions when idle energy consumption is significant, so that a Power-Down ratio, β, is defined as:

$$\beta = \frac{Power\ dissipation\ while\ idling}{Power\ dissipation\ while\ computing} = \frac{P_{IDLE}}{P_{MAX}} \quad \text{(EQ 2.25)}$$

E_{IDLE} can now be expressed as a function of E_{MAX}:

$$E_{IDLE} = \frac{[\beta \cdot E_{MAX} T_{MAX}] \cdot [(1 - T_{AVE}/T_{MAX})t_S]}{[T_{AVE}] \cdot [t_S]} \quad \text{(EQ 2.26)}$$

Equation 2.27 shows that idle energy consumption dominates total energy consumption when the fractional time spent computing (T_{AVE}/T_{MAX}) is less than the fractional power dissipation while idling (β).

$$BETR = ETR\left[1 + \beta\left(\frac{T_{MAX}}{T_{AVE}} - 1\right)\right], \quad T_{MAX} \geq T_{AVE} \quad \text{(EQ 2.27)}$$

To show that *BETR* represents a general metric for energy use taking into account the throughput and idle energy, the limits of the *BETR* metric will be explored.

Idle Energy Consumption is Negligible ($\beta \ll T_{AVE}/T_{MAX}$): The metric should simplify to that found in the maximum throughput mode, since it is only during the bursts of computation that energy is consumed and operations performed. For negligible power dissipation during idle, the *BETR* metric in Equation 2.27 degenerates to the *ETR*, as expected. For perfect power-down ($\beta = 0$) or high user demanded throughput ($T_{MAX} = T_{AVE}$), the *BETR* is exactly the *ETR*.

Idle Energy Consumption Dominates ($\beta \gg T_{AVE}/T_{MAX}$): The energy efficiency should increase by either reducing the idle energy/operation while maintaining constant throughput, or by increasing the throughput while keeping idle energy/operation constant. While it might be expected that these are independent optimizations, E_{IDLE} may be related back to E_{MAX} and the throughput by β since T_{AVE} is fixed:

$$\frac{E_{IDLE}}{E_{MAX}} \cong \frac{P_{IDLE}/T_{AVE}}{P_{MAX}/T_{MAX}} = \beta \cdot \frac{T_{MAX}}{T_{AVE}} \quad \text{(EQ 2.28)}$$

Expressing E_{IDLE} as a function of E_{MAX} yields:

$$BETR \cong \frac{\beta \cdot E_{MAX}}{T_{AVE}}, \quad \text{(EQ 2.29)}$$

when the idle energy dominates. If β remains constant for varying throughput (and E_{MAX} stays constant), then E_{IDLE} scales with throughput as shown in Equation 2.28. Thus, the *BETR* becomes an energy/operation minimization similar to the fixed throughput mode. However, β may vary with throughput since the energy to go in and out of the idle condition may be much larger than the steady-state condition as well as the possibility of different levels of the idle state as discussed in Section 3.2.6.

2.3.4 Energy Efficiency for Practical Designs

Even though, the *BETR* metric measures the energy use of general purpose processor systems, the information on the system's average throughput (T_{AVE}) is required to utilize this metric is application specific. Thus, the *BETR* metric cannot be used to describe the energy efficiency of a processor in general terms, but requires the specification of a target application, or class of related applications. An example application is the InfoPad, as described in Section 2.1.3, in which the processor system is responsible for packet-level network control on the pad and has an average throughput requirement of 0.8 MIPS. If the video decompression was implemented by the processor rather than the custom chip-set, then the average throughput would increase to approximately 11 MIPS, a very different optimization problem. However, the *BETR* metric's subcomponents, *ETR* and E_{IDLE} can be investigated in an application independent way and this will be done in the following section to motivate some general principles of energy efficient design.

2.4 Energy Efficient Design Observations

Several observations can be made if the previously defined metrics are used to minimize the energy required to provide a given level throughput: design for high speed then slow down the circuit using voltage reduction; operate as low a speed as the application will allow; don't use clock frequency reduction to improve energy efficiency; and dynamic voltage scaling provides a strategy to give the appearance of high throughput operation with the energy efficiency of a low speed design.

2.4.1 Design for speed then slow it down

If a processor is designed to have a high clock rate and if the application requirement does not require the maximum performance achieved, the "excess" speed can be used to improve energy efficiency. This is accomplished by reducing the supply voltage so that the critical path is set to just meet the application requirements. This is such an effective approach that processors which weren't designed to be energy efficient, can in fact can be made efficient if their supplies are reduced, even in comparison to processors explicitly used in low power situations. There are, however, techniques that are so inefficient that the energy reduction that are achieved cannot be compensated for with supply reduction, so this approach to energy efficient design must be carefully applied.

As an example of the importance of supply reduction a comparison of two processors will be made. Table 2.1 lists the characteristics of these processors – the ARM710 which targets the low-power market, and the R4700 which targets the mid-range workstation market, with both being fabricated in similar 0.6μm technologies. The measure of throughput used is SPECint92. A commonly-used metric for measuring energy efficiency is SPECint92/Watt (or SPECint95/Watt, Dhrystones/Watt, MIPS/Watt, etc.). The ARM710 processor has a SPECint92/Watt five times greater than the R4700's, and the claim then follows that it is "five times as energy efficient".

TABLE 2.1 Comparison of two commercial processors *[2.23][2.24]*.

Processor	SPECint92 (T_{MAX})	Power (Watts)	Supply voltage, V_{DD} (Volts)	SPECint92/Watt ($1/E_{MAX}$)	ETR (10^{-3})
R4700	130	4.0	3.3	33	0.24
ARM710	20	0.12	3.3	167	0.30

Energy Efficient Design Observations

However, this metric only compares operations/energy, and does not weight the fact that the ARM710 has only 15% of the performance as measured by SPECint92. The ETR (Watts/SPECint92^2) metric which does incorporate this throughput disparity indicates that the R4700 is actually 25% more energy efficient in providing a given level of throughput than the ARM710 (.24 vs. .30). To quantify the efficiency increase, the plot of energy/operation versus throughput in Figure 2.12 is used because it better tracks the R4700's energy at the low throughput values. The plot was generated from the throughput and energy/operation models in Section 2.2.

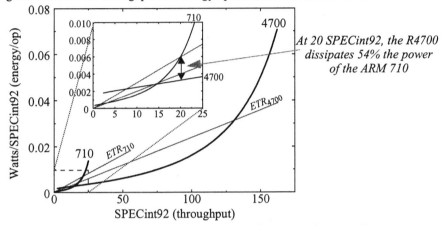

FIGURE 2.12 Energy vs. Throughput of R4700 and ARM710.

According to the plot, the R4700 would dissipate 65mW at 20 SPECint92, or about 1/2 of the ARM710's power, despite the low V_{DD} ($1.5 \cdot V_T$) for the R4700. Conversely, the R4700 can deliver 30 SPECint92 at 120mW ($V_{DD} = 1.7 \cdot V_T$), or 150% of the ARM710's throughput.

This does assume that the R4700 processor has been designed so that it can operate at these low supply voltages. If the lower bound on operating voltage is greater than $1.7 \cdot V_T$, then the ARM710 would be more energy efficient in delivering the 20 SPECint92 than the R4700. Typically, a processor is rated for a fixed standard supply voltage (3.3V or 5.0V) with a ±10% tolerance. However, many processors can operate over a much larger range of supply voltages (e.g., 2.7-5.5V for the ARM710 [2.23], 2.0-3.3V for the Intel486GX [2.25]). The processor can operate at a non-standard supply voltage by using a high-efficiency, low-voltage DC-DC converter to generate the appropriate supply voltage [2.26]

Energy Efficient Design

While the *ETR* correctly predicted the more energy-efficient processor at 20 SPECint92, it is important to note that the R4700 is not more energy efficient for all values of SPECint92. Because the nominal throughput of the processors is vastly different, the Energy/Operation versus Throughput plot better tracks the efficiency than the implicit linear approximation used in the ETR. This plot shows a cross-over throughput of 14.5 SPECint92, at which below this value, the ARM710 becomes more efficient indicating the importance of proper usage modeling for the best results.

2.4.2 Operate at as low a speed as possible

A related principle is that the processor should be run as slow as possible, in order to exploit the dual advantages of an improvement in energy efficiency which results from voltage reduction and to minimize wasted energy resulting from idling. This is an alternative to the typical situation which provides the user the fastest possible response time by operating the processor at the highest possible supply voltage, and then shuts down (if possible) between the activity periods. While this approach provides the user with the minimum latency, it is a particularly inefficient approach if the user requires burst mode operation with a low duty cycle.

For example, assume the target application has a T_{AVE} of 20 SPECint92, and both the ARM710 and R4700 have a β (power down ratio) of 0.2. If both processors' V_{DD} is left at 3.3V, The ARM710's *BETR* is exactly equal to its *ETR* value, which is 3.0×10^{-4}. It remains the same because it never idles. The R4700, on the other hand, spends 85% ($T_{AVE}/T_{MAX}=.15$) of the time idling, and its *BETR* is 5.0×10^{-4}. Thus, for this scenario, the ARM710 is nearly twice as energy efficient.

However, if the R4700's β can be reduced down to 0.02, then the *BETR* of the R4700 becomes 2.66×10^{-4}, and it is once again the more energy-efficient solution. For this example, the cross-over value of β is 0.045.

This example demonstrates how important it is to use the *BETR* metric instead of the *ETR* metric if the target application's idle time is significant (i.e., T_{AVE} can be characterized and is significantly below T_{MAX}). For the above example, a β for the R4700 greater than 0.045 leads the metrics to disagree on which is the more energy-efficient solution. One might argue that the supply voltage can always be reduced on the R4700 so that it is more energy efficient for any required throughput. This is true if the dynamic range of the R4700 is as indicated in Figure 2.12. However, if some internal logic limited the value that V_{DD} could be dropped, then the lower bound on the R4700's throughput could be located at a much higher value. Thus, finite β can degrade the energy efficiency of a high-throughput processor, due to excessive idle power dissipation.

2.4.3 Slowing the clock without voltage reduction wastes energy

A common fallacy is that by simply reducing the clock frequency, f_{CLK}, there will be energy savings. Reducing f_{CLK} does reduce *power* dissipation, but it does not increase energy efficiency and in fact when energy consumption due to active computation dominates idle energy consumption, clock rate reduction actually decreases the energy use efficiency, ETR. At best, when idle energy consumption is dominant, it allows an energy-throughput trade-off.

The relative amount of time spent idling versus computing is an important consideration in determining the effect of clock frequency reduction on energy efficiency. A couple of limits will be explored.

Compute energy consumption dominates ($E_{MAX} \gg E_{IDLE}$): Since compute energy consumption is independent of f_{CLK}, and throughput scales proportionally with f_{CLK}, decreasing the clock frequency increases the *ETR*, and thereby reduces energy efficiency. Halving f_{CLK} is equivalent to doubling the computation time, while maintaining constant computation per battery life, which is clearly energy inefficient.

Idle energy consumption dominates ($E_{IDLE} \gg E_{MAX}$): Clock reduction may trade-off throughput and energy/operation, but only when the power-down efficiency, β, is independent of throughput such that E_{IDLE} scales with throughput. When this is so, halving f_{CLK} will double the computation time, but will also double the amount of computation per battery life, since E_{IDLE} has been halved. If the currently executing process can tolerate throughput degradation, then this may be a reasonable trade-off. If β is inversely proportional to throughput, however, then reducing f_{CLK} does not affect the total energy consumption, and the energy efficiency drops.

As shown in Table 2.2, reducing the clock frequency increases energy use in two of the three possible operating conditions. In the third operating condition, the efficiency

Energy Efficient Design

merely remains unchanged. Thus, simple clock frequency reduction is never energy efficient.

TABLE 2.2 Impact of Clock Frequency Reduction on Energy Use.

Operating Conditions:	Compute Energy Consumption Dominates	Idle Energy Consumption Dominates	
		β (powerdown ratio) independent of throughput	β inversely proportional to throughput
Throughput	decreases	decreases	decreases
Energy	unchanged	decreases	unchanged
Energy to throughput ratio, $BETR$	increases	unchanged	increases

2.4.4 Varying both the supply voltage and throughput is optimal

The observations above indicate that it is desirable to operate at as low a clock rate (throughput) as possible, but that unfortunately is often unacceptable if the user is then required to suffer increased latency in their requests for computation, however rare they may be. In order to provide the energy efficiency that is achievable at low clock rates, with out a penalty in response time it is necessary to dynamically vary the clock rate so that the high throughput mode is only provided at those times that the user can detect it. In other words the goal is to give the user the appearance of a high throughput operation, even though most of the time the processor is operating in the much more energy efficient low throughput mode.

To achieve the energy efficiency in the low throughput mode it is necessary to require V_{DD} to track f_{CLK}, so that the critical path delay just meets the required clock frequency requirement. This is equivalent to the V_{DD} scaling of Section 2.3.2 except that it is done dynamically during processor operation. If there is significant idle time, so that E_{IDLE} dominates the total energy consumption, there will be the largest opportunities for improvement in energy efficiency.

However, even when idle energy consumption is negligible, dynamically varying the supply voltage (which will be referred to as Dynamic Voltage Scaling or DVS) provides significant wins. Figure 2.13 plots a sample usage pattern of desired throughput, with the delivered throughput super-imposed on top. For background and high-latency tasks, the supply voltage can be reduced so that just enough throughput is delivered, which minimizes energy consumption.

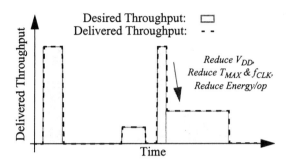

FIGURE 2.13 Dynamic Voltage Scaling.

For applications that require maximum deliverable throughput only a small fraction of the time, dynamic voltage scaling provides a significant energy efficiency improvement. For the R4700 processor, the peak throughput is 130 SPECint92. Given a target application where the desired throughput is either a fast 130 SPECint92 or a slow 13 SPECint92, Table 2.3 lists the peak throughput, average energy/operation, and effective *ETR* depending on the fraction of time spent in the fast mode.

If the total number of operations are the same for different modes of use the relative changes in energy consumption (battery life) can be evenly compared. When the fraction of time in maximum throughput mode becomes small, the processor's peak throughput is still set by the fast (high voltage) mode, while the average energy consumed per operation is set by the slower (low voltage) mode. Thus, the best of both worlds can be achieved - high performance at low average energy consumption

TABLE 2.3 Benefits of Dynamic Voltage Scaling.

Throughput:	Time spent operating in:			T_{MAX} (SPECint92)	E_{MAX} (SPECint92)	ETR (10^{-6})	Normalized Battery Life
	Fast Mode	Slow mode	Idle Mode				
Always full-speed	10%	0%	90%	130	0.031	237	1 hr.

Energy Efficient Design

TABLE 2.3 Benefits of Dynamic Voltage Scaling.

Throughput:	Time spent operating in:			T_{MAX} (SPEC int92)	E_{MAX} (SPEC int92)	ETR (10^{-6})	Normalized Battery Life
	Fast Mode	Slow mode	Idle Mode				
Sometimes full-speed	1%	90%	9%	130	0.006	45.0	5.3 hrs.
Rarely full-speed	0.1%	99%	0.9%	130	0.003	25.8	9.2 hrs.

As shown in Table 2.3, the battery run-time can be improved by up to a factor of 10x. In most portable devices (e.g. notebook computers, PDAs, etc.), peak throughput is typically used only a small fraction of the time, such that this energy-efficiency improvement is readily achievable. Although dynamically varying V_{DD} and f_{CLK} in a processor system may seem extraordinarily difficult to accomplish, Chapter 4 will demonstrate that dynamic voltage scaling is a relatively straightforward and simple technique to implement.

There are three key components for implementing DVS in a general-purpose microprocessor system: an operating system that can intelligently vary the processor speed, a regulation loop that can generate the minimum voltage required for the desired speed, and a microprocessor that can operate over a wide voltage range.

The next section focuses on the implementation methodology, the new system constraints, and the impact of DVS on the hardware design methodology. This methodology was validated by a prototype embedded processor system that successfully implements DVS, which is described fully in Chapter 6.

2.5 Dynamic Voltage Scaling

The unique advantages of dynamic variation of the supply voltage in providing a new degree of freedom in improving energy efficiency would not be interesting if it adds considerable effort to the design task. Fortunately, Digital CMOS circuits are very amenable to implementing DVS, as their performance and energy consumption can easily be made to scale over a wide range of supply voltage. Although the maximum supply voltage drops with improved process technology, thereby reducing this range, so does the device threshold voltage, such that DVS will continue to be a viable technique for future process technologies. Furthermore, DVS can also provide a solution to the leakage problem of low threshold-voltage processes by scaling leakage current with supply voltage.

2.5.1 Voltage Scaling Effects on Circuit Delay

CMOS circuit delay tracks very well over supply voltage, as shown in Figure 2.14. Four example circuits are shown, ranging in complexity from a simple inverter to a complex SRAM design that consists of an address decoder, memory cell array, sense amplifier, output buffer, and control sequencing logic. The maximum clock speed is just the inverse of the critical path delay, which was calculated via a SPICE simulation and then normalized at 4V (the SPICE data for the SRAM is from *[2.27]*).

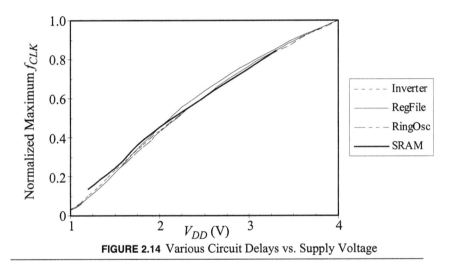

FIGURE 2.14 Various Circuit Delays vs. Supply Voltage

The inverter, ring oscillator, and register file all vary less than 10% over the full range of supply voltage, V_{DD}. These three circuits are a good cross-representation of the bulk of CMOS circuits, both in logic style and complexity. The SRAM circuit, which differs from the others because NMOS devices dominate its critical path delay, runs faster at lower voltage because for our 0.6μm process, $V_{Tn} < V_{Tp}$. However, the speed variation at 1.2V is still only 25%, which is insignificant compared to the 10x overall reduction in maximum speed from 4V. More importantly, the deviation is in the positive direction; because the SRAM circuit runs faster at lower V_{DD}, it will not be a limiting factor of the chip's speed at low V_{DD}.

Energy Efficient Design

Energy Efficient Design

By using a ring oscillator to generate the clock signal, the clock frequency can be scaled lock-step with V_{DD}, enabling proper operation of the processor over the full range of V_{DD} through a closed-loop control system.

2.5.2 Maximum Energy Efficiency Improvement.

Figure 2.15 demonstrates the possible energy-efficiency improvement of DVS. Starting at the nominal V_{DD} operating point of 3.3V, when the clock frequency, f_{CLK}, is reduced, there is a proportional decrease in throughput. When this is done at constant V_{DD}, there is no reduction in energy/operation. However, if V_{DD} is scaled lock-step with f_{CLK}, then the lower curve is traversed, yielding more than a 10x energy reduction at low voltage.

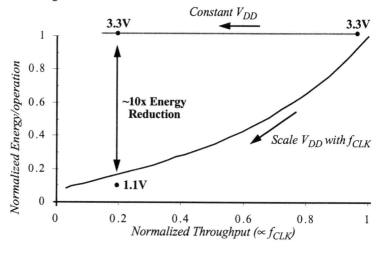

FIGURE 2.15 Scaling V_{DD} with f_{CLK}.

The ability to dynamically traverse this curve is how DVS can improve energy efficiency. For the processor described in Chapter 6, the lower operating point is 6 MIPS @ 0.27 mW/MIPS ($ETR = 45$ µW/MIPS^2), and the upper operating point is 85 MIPS @ 2.8 mW/MIPS ($ETR = 33$ µW/MIPS^2). However, if peak throughput is only occasionally demanded, then the processor can deliver a peak throughput of 85 MIPS, while the average energy/operation can be as low as 0.27 mW/MIPS. This yields an ETR of 3.2 µW/MIPS^2, which is more than a 10x improvement in the energy required to sustain this level of apparent throughput.

Figure 2.16 plots the normalized battery run-time, which is inversely proportional to energy/operation, as a function of the fractional amount of computation performed at low throughput for the above processor. While a moderate run-time increase (22%) can be achieved with only 20% of the computation at low throughput, DVS yields significant increases when more of the computation can be run at low throughput, with the upper limit in excess of a 10x increase in battery run-time, or equivalently, more than a 10x reduction in energy/operation.

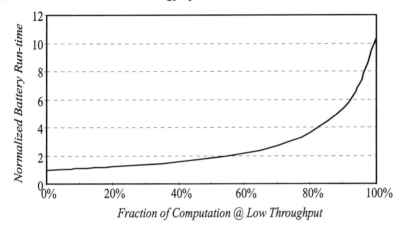

FIGURE 2.16 Battery Run-time vs. Workload Ratio

2.5.3 Essential Components

A typical processor system is powered by a voltage regulator which outputs a fixed voltage. However, the implementation of DVS requires a voltage converter that can dynamically adjust its output voltage when requested by the processor to do so. In Chapter 7 the design of such a converter is described which can be integrated in CMOS [2.28]. This functionality can also be implemented with discrete commercial components, but with much lower conversion efficiency than an optimized design.

Another essential component is a ring oscillator matched to the processor's critical paths, such that as the critical paths vary over V_{DD}, so will the processor clock frequency. This is best achieved by having a ring oscillator on the processor, which will then track the critical paths over process and temperature.

Energy Efficient Design

The processor itself must be designed to operate over the full range of voltage supply, which places restrictions on the types of circuits that can be used and impacts processor verification, as described in Section 5.6. Additionally, the processor must be able to properly operate while V_{DD} is varying, as detailed in Section 4.5.

The last essential component is a DVS-aware operating system. The hardware itself has no knowledge of the priority of the currently executing task, since this information only resides within the operating system scheduler. Hence, to deliver the significant increase in energy efficiency afforded by DVS, the operating system must be able to intelligently vary V_{DD} and f_{CLK} as a function of desired throughput, which is further described in Chapter 10.

2.5.4 Fundamental Trade-Off

Processors generally operate at a fixed voltage, and require a regulator to tightly control voltage supply variation. The processor produces large current spikes for which the regulator's output capacitor supplies the charge. Hence, a large output capacitor on the regulator is desirable to minimize ripple on V_{DD}. A large capacitor also helps to maximize the regulator's conversion efficiency by reducing the voltage variation at the output of the regulator.

However, the voltage converter required for DVS is fundamentally different from a standard voltage regulator because in addition to regulating voltage for a given clock frequency, it must also change the operating voltage when a new clock frequency is requested. To minimize the speed and energy consumption of this voltage transition, a small output capacitor on the converter is desirable, in contrast to the supply ripple requirements.

Thus, the fundamental trade-off in a DVS system is between good voltage regulation and fast/efficient dynamic voltage conversion. As will be shown in Chapter 7, it is possible to optimize the size of this capacitor to balance the requirements for good voltage regulation with the requirements for a good dynamic voltage conversion.

2.5.5 Scalability with Technology

For DVS to provide significant energy efficiency improvement, the process technology must be able to operate over a wide range of voltage, such that the throughput and energy consumption can vary significantly to allow for increased opportunities for energy reduction.

The lower bound on voltage is set by the larger of V_{Tn} and V_{Tp}, beyond which the MOSFET's begin operating in the subthreshold region, and their delay increases exponentially [2.20]. A more practical limit is ~100mV above $\max(V_{Tp}, V_{Tn})$, to provide an operating margin for preventing the MOSFET's from entering this region.

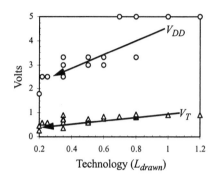
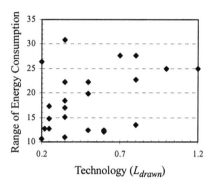

FIGURE 2.17 V_{DD}, V_T, and Range of Energy Consumption vs. Process Technology.

The upper bound on voltage is determined by gate-oxide breakdown [2.13]. For our 0.6μm process, this is only 6.3V. To provide a margin of safety, a process has a rated maximum voltage of around one-half of the gate-oxide breakdown voltage; for the 0.6μm process, the rated maximum voltage is 3.3V. While the MOSFET's can be operated at a higher voltage, it is generally not recommended for long-term gate-oxide reliability.

As process technology advances, the reduction in gate-oxide thickness necessitates a reduction in the rated maximum supply voltage. However, to maintain MOSFET performance, their threshold voltages have also been reduced, as shown by the sampling of 24 process technologies in Figure 2.17. Scaling V_T with V_{DD} maintains a large range of energy consumption ($V_{DD}^2 / (V_T + 100\text{mV})^2$), anywhere from 10x to over 30x. A future 0.10μm process may only have a rated voltage of 1.2V, but with a V_T of 0.3V and 50mV of operating margin, the possible energy range is still 11.8x. Since throughput scales by a similar order of magnitude as does energy consumption, DVS is still quite applicable to even deep-submicron process technologies.

References

[2.1] Standard Performance Evaluation Corporation, *SPEC Run and Reporting Rules for CPU95 Suites*, Technical Document, Sept. 1994.

[2.2] A. Chandrakasan, S. Sheng, and R.W. Brodersen, "Low-Power CMOS Digital Design", *IEEE Journal of Solid State Circuits*, Apr. 1992, pp. 473-84.

[2.3] A. Chandrakasan, R.W. Brodersen, *Low-power Digital CMOS Design*, Kluwer Academic Publishers, Boston, 1995.

[2.4] M. Culbert, "Low Power Hardware for a High Performance PDA", *Proceedings of the Thirty-Ninth IEEE Computer Society International Conference*, Mar. 1994, pp. 144-7.

[2.5] T. Ikeda, "ThinkPad Low-Power Evolution", *Proceedings of the IEEE Symposium on Low Power Electronics*, Oct. 1995, pp. 6-7.

[2.6] A. Chandrakasan, A. Burstein, and R.W. Brodersen, "A Low Power Chipset for Portable Multimedia Applications", *IEEE Journal of Solid State Circuits*, Vol. 29, No. 12, Dec. 1994, pp. 1415-28.

[2.7] S. Kunii, "Means of Realizing Long Battery Life in Portable PCs", *Proceedings of the IEEE Symposium on Low Power Electronics*, Oct. 1995, pp. 20-3.

[2.8] GEC Plessey Semiconductor, *ARM60 Data Sheet*, Technical Document, Aug 1994.

[2.9] Digital Equipment Corporation, *DIGITAL Semiconductor SA-1100 Microprocessor Technical Reference Manual*, Document EC-R5MTB-TE, Jan. 1998.

[2.10] J. Rabaey, *Digital Integrated Circuits, A Design Perspective*, Prentice Hall, Upper Saddle River, NJ, 1996.

[2.11] P. Landman, J. Rabaey, "Black-Box Capacitance Models for Architectural Power Analysis", *Proceedings of the 1994 International Workshop on Low-Power Design*, Napa Valley, CA, April 1994.

[2.12] H. Veendrick, "Short-Circuit Dissipation of Static CMOS Circuitry and Its Impact on the Design of Buffer Circuits", *IEEE Journal of Solid State Circuits*, Vol. 19, No. 4, August 1984.

[2.13] R. Muller, T. Kamins, *Device Electronics for Integrated Circuits*, Wiley, New York, 1986.

[2.14] S. Sze, *Physics of Semiconductor Devices*, Wiley, New York, 1981.

[2.15] J. Montanaro, et. al., "A 160-MHz 32-b 0.5-W CMOS RISC Microprocessor", *IEEE Journal of Solid State Circuits*, Vol. 31, No. 11, Nov. 1996, pp. 1703-14.

[2.16] V. De and S. Borkar, "Technology and Design Challenges for Low Power and High Performance", *Proceedings of the IEEE Symposium on Low Power Electronics and Design*, Aug. 1999, pp. 163-8.

[2.17] S. Mutoh, et. al., "1-V Power Supply High-Speed Digital Circuit Technology with Multithreshold-Voltage CMOS", *IEEE Journal of Solid State Circuits*, Vol. 30, No. 8, Aug. 1995, pp. 847-54.

[2.18] T. Kuroda, et. al., "Variable Supply-voltage Scheme for Low-power High-speed CMOS Digital Design", *IEEE Journal of Solid State Circuits*, Vol. 33, No. 3, Mar. 1998, pp. 454-62.

[2.19] K. Toh, P. Ko, R. Meyer, "An Engineering Model for Short-Channel MOS Devices", *IEEE Journal of Solid-State Circuits*, Vol. 23, No. 4, April 1988.

[2.20] R. Pierret, *Semiconductor Device Fundamentals*, Addison Wesley, Reading, MA, 1996.

[2.21] J. Huang, et. al., "A Robust Physical and Predictive Model for Deep-Submicrometer MOS Circuit Simulation", *Proceedings of the IEEE Custom Integrated Circuits Conference*, May 1993, pp. 14.2.1-4.

[2.22] M. Horowitz, T. Indermaur, and R. Gonzalez, "Low-Power Digital Design", *Proceedings of the IEEE Symposium on Low Power Electronics*, Oct. 1994, pp. 8-11.

[2.23] Advanced RISC Machines, Ltd., *ARM710 Data Sheet*, Technical Document, Dec. 1994.

[2.24] Integrated Device Technology, Inc., *Enhanced Orion 64-Bit RISC Microprocessor*, Data Sheet, Sept. 1995.

[2.25] Intel Corp., *Embedded Ultra-Low Power Intel486TM GX Processor*, SmartDieTM Product Specification, Dec. 1995.

[2.26] A. Stratakos, S. Sanders, and R.W. Brodersen, "A Low-voltage CMOS DC-DC Converter for Portable Battery-operated Systems", *Proceedings of the Twenty-Fifth IEEE Power Electronics Specialist Conference*, June 1994, pp. 619-626.

[2.27] A. Burstein, *Speech Recognition for Portable Multimedia Terminals*, Ph.D. Thesis, University of California, Berkeley, Document No. UCB/ERL M97/14, 1997.

[2.28] A. Stratakos, *High-Efficiency, Low-Voltage DC-DC Conversion for Portable Applications*, Ph.D. Thesis, University of California, Berkeley, 1998.

CHAPTER 3 *Microprocessor System Architecture*

While it is important to be aware of energy issues at all levels of the design hierarchy, energy-efficiency optimizations at the level of architectural design generally yield the largest gains. The closer the design approaches to the final physical implementation, the more difficult the gains become because the scope of possible optimizations narrows.

Unfortunately, traditional architectural design methodologies for microprocessor systems focus primarily on performance, with energy consumption considered as an afterthought. This approach is slowly evolving to consider energy consumption earlier in the design process, but a radical change in design methodology that is required has yet to happen. This wholesale change requires the incorporation of energy estimation into the high-level system simulator (Section 5.2), so that both performance and energy consumption can be estimated at the highest level of the design space. This allows for architectural design choices to be evaluated for their overall impact on system energy efficiency, and not just strictly performance.

The first section describes the architectural design and methodology of the processor system, while subsequent sections describe in more detail the major components of the microprocessor itself -- the processor core, the cache system, and the system-control coprocessor.

3.1 System Architecture

Microprocessor systems generally resemble the generic architecture shown in Figure 3.1. The processor bus connects the microprocessor to the main external memory (ROM, RAM), input/output devices via I/O controllers, and peripheral subsystems via bus controllers. A PAL or PLD is typically used to generate the control signals between the various chips.

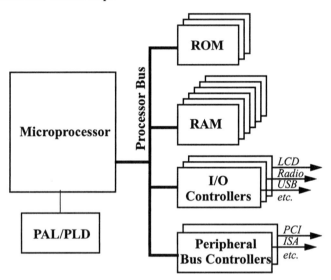

FIGURE 3.1 Generic Microprocessor System Architecture.

While the generic system appears to have little room for optimization, a number of architectural design choices are available which can significantly impact performance and energy consumption.

3.1.1 Main Memory Architecture & Processor Bus Topology

Commodity SRAM's and DRAM's used for main memory are typically eight bits wide. This data width has been used since the earliest microprocessor days when the processor itself was only eight bits wide.

System Architecture

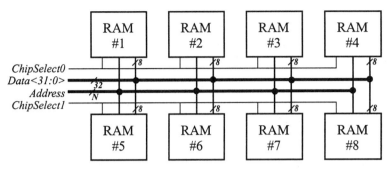

FIGURE 3.2 Typical Main Memory Architecture.

While the bus width of the microprocessor has increased to improve memory bandwidth to the processor, the width of the RAM chips has remained unchanged. To increase the bandwidth of the memory bank, multiple memory chips are enabled in parallel, as shown in Figure 3.2. In this example, four 8-bit RAM chips are accessed simultaneously to provide a 32-bit data word to the processor. While this approach successfully meets the bandwidth (performance) demand, it increases the energy consumed per access. However, a 32-bit RAM chip would require only one memory chip to be activated per access, eliminating the unnecessary energy consumption of the other three chips. The primary drawback to a larger bus width is increased pin count on the memory chip. A typical 8-bit wide, 16Mb RAM chip has a total of 29 address and data pins. However, if the address and data are multiplexed onto the same bus, then only five additional pins are required to support 32-bit accesses as shown in Figure 3.3.

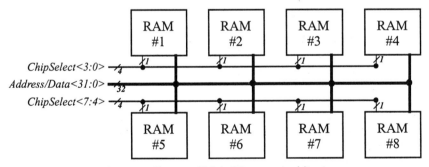

FIGURE 3.3 Proposed Main Memory Architecture.

Microprocessor System Architecture   **47**

Two bus cycles are required for a single-word memory access in this multiplexed approach. However, for a processor with a cache, most memory accesses are cache line reloads. Thus, if the processor bus can support burst-mode accesses, then the cost of placing the address on the multiplexed bus can be amortized over multiple data words. For a cache-line length of eight, the effective bandwidth is 8/9, or 89% of peak capacity. Benchmark simulations (the benchmarks are described in Section 10.4 and the simulation strategy is described in Section 5.2) demonstrated that with a large 16kB cache, external accesses are predominantly cache-line reads and writes such that actual bus capacity is only 15% less than peak capacity. This slight degradation in performance is more than compensated for by the reduction in energy consumption of the external memory chips. Thus, the processor bus was designed to support burst accesses, as described in further detail in Section 6.3.1.

3.1.2 I/O Considerations

The I/O interface chip is essentially a bridge between the processor bus and the I/O bus, such that its implementation is relatively simple. The critical functionality required is flow control between the processor and the autonomous I/O controllers, which all operate at different clock frequencies. This is provided through a simple state machine on the chip in conjunction with wait signals which can halt either the processor, or an I/O controller, as necessary.

To improve the system energy efficiency, the I/O interface chip also allows for packed I/O data writes. Typical I/O data transfers are one byte wide, such that if I/O data is transferred in single bytes, two processor bus cycles are required to complete the data transfer, one for the address and one for the data byte. The aggregate I/O data transfer rate is 0.5 bytes/cycle. However, by allowing four bytes to be packed into a single word, then the transfer rate increases to 2 bytes/cycle. The I/O interface chip is responsible for unpacking the word into four individual byte writes to the addressed I/O controller. Packing I/O byte reads is not done, since the processor would have to wait until the fourth read before it could process the data, adding additional latency to the I/O transfer.

To improve system performance, the I/O interface chip also supports direct memory accesses (DMA) to allow I/O controllers to directly access the main memory. This frees up processor cycles that would otherwise be required since the processor must act as an intermediary between the I/O controllers and main memory when DMA is not supported.

In the prototype system (Chapter 9), this functionality is performed by the virtual I/O controller (Section 9.3) which eliminates the need to implement this added functionality in silicon. The actual I/O interface chip that was implemented only provides the voltage level-conversion and signalling required to support I/O transfers and DMA between the processor bus and the virtual I/O controller. To improve flexibility in testing and system evaluation this controller was implemented with a Xilinx FPGA and a StrongArm processor system. For a complete system solution, this functionality must be implemented on the I/O interface chip, however since the average number of I/O accesses per processor cycle is well below one, the energy consumed by these I/O operations in a custom implementation could easily be designed to be insignificant.

3.2 Processor Core

The design of the processor core presents a variety of opportunities for improving the overall microprocessor system's energy efficiency. To facilitate the design of the prototype low energy processor of Chapter 6, an ARM8 behavioral model from ARM Ltd. was incorporated into the high-level design specification of the prototype processor. While significantly speeding up the implementation, this model also constrained the processor design space by fixing the instruction set architecture as well as the core microarchitecture.

As such, this section explores common processor core optimizations for improving performance and/or reducing power dissipation. To properly evaluate these optimizations, it is imperative to analyze the system-level performance and energy consumption improvement, rather than focus only on the improvement within an individual block.

3.2.1 Instruction Set Architecture

Typically, an instruction set architecture (ISA) is designed solely with performance in mind. High-level performance simulators allow the architect to explore the ISA design space with reasonable efficiency. Energy is not a consideration, nor are there high-level simulators available to even let the architect estimate energy consumption. Simulation tools exist, but require a detailed description of the microarchitecture so that they are not useful until the ISA has been completely specified. Processors targeted towards portable systems should have their ISA designed for energy efficiency, and not just performance.

Microprocessor System Architecture

Instruction word length

Many processors have 32-bit instruction-words and registers. Register width generally depends on the required memory address space, and cannot be reduced; in fact, more recent microprocessors have moved to 64 bits. On the other hand, for low-energy processors, 16-bit instruction widths have been considered to be more appropriate. Static code density can be reduced by 30-35%, while increasing the dynamic run length by only 15-20% over an equivalent 32-bit processor*[3.1]][3.2]*. Using 16-bit instructions reduces the energy cost of an instruction fetch by up to 50% because the size of the memory read has been halved *[3.3]*. In system's with 16-bit external busses, the advantage of 16-bit instructions is further widened*[3.4]*. Since instruction fetch consumes about a third of the processor's energy*[3.5][3.6]*, total energy consumption is reduced by 15-20% but since there can be a 15-20% reduction in performance due to reformatting for a 32 bit external bus, the ETR is unchanged.

However, instruction word length reduction can improve energy efficiency if the external memory energy consumption dominates the processor energy consumption. This would only be true for very small, on-chip caches (< 8kB). Since most microprocessors today contain at least a moderately-sized on-chip cache (16kB or larger) enabled by CMOS VLSI process scaling, utilizing 16-bit instructions will typically have negligible impact on microprocessor energy efficiency. It is likely only to be useful for reducing memory requirements in cost-sensitive system designs or in processors with a reduced bus sizes.

Register file size

The register file consumes a sizable fraction of total energy consumption since it is typically accessed multiple times per cycle (10% of the total energy in *[3.5]*, as well as in the prototype design of Chapter 6). In a register-memory architecture, the number of general purpose registers is kept small and many operands are fetched from memory. Since the energy cost of a cache access surpasses that of a moderately sized (32) register file, this is not an energy efficient solution. The other extreme is to implement register windows which is essentially a very large (100+) register file. In this case the energy consumed by the register file increases dramatically increasing total processor energy consumption by as much as 10-20%, requiring an significant, probably unrealizable, increase in performance if ETR is to be improved *[3.7]*.

CISC vs.RISC

The issue of supported operation types and addressing modes has been a main philosophical division between the RISC and CISC proponents. While this issue has been debated solely in the context of performance, it can also have an impact on energy consumption. Complex ISA's have higher code density, which reduces the energy consumed fetching instructions and reduces the total number of instructions executed. Simple ISAs typically have simpler data and control paths, which reduces the energy consumed per instruction, but there are more instructions. These trade-offs need to be analyzed when creating an ISA.

Compiler optimizations

The amount of hardware exposed (e.g., branch delay slot, load delay slot, etc.) is another main consideration in ISA design. This is typically done to improve performance by simplifying the hardware implementation. Since the scheduling complexity resides in the compiler, it consumes zero run-time energy while the simplified hardware consumes less energy per operation. Thus, both the performance is increased and the energy/operation is decreased, giving a two-fold increase in ETR. A good example of radically exposing the hardware architecture are very long instruction word (VLIW) architectures, which will be discussed in more detail in the following section.

3.2.2 Architectural Concurrency

The predominant technique to increase energy efficiency in custom DSP IC's (fixed throughput) is architectural concurrency; with regards to processors, this is generally known as instruction-level parallelism (ILP). Previous work on fixed throughput applications demonstrated an energy-efficiency improvement of approximately N on an N-way parallel/pipelined architecture *[3.8]*. This assumes that the instructions being executed are fully vectorizable, that N is not excessively large, and that the extra delay and energy overhead for multiplexing and demultiplexing the data is insignificant.

Moderate pipelining (4 or 5 stages), while originally implemented purely for speed, also increases energy efficiency, particularly in RISC processors which operate near one cycle-per-instruction. ETR can be improved by a factor of two or more*[3.9]*, and is thus essential in an energy-efficient processor.

Superscalar Architectures

More recent processor designs have implemented superscalar architectures with parallel execution units, in the hope of further increasing the processor's execution concurrency. However, an N-way superscalar machine will not yield a speedup of N, due

Microprocessor System Architecture

to the limited ILP found in typical code *[3.10][3.11]*. Therefore, the achievable speedup will be less than the number of simultaneous issuable instructions and yields diminishing returns as the peak issue rate is increased. The speedup has been shown to be between two and three for practical hardware implementations *[3.12]*.

If the instructions are dynamically scheduled in employing superscalar operation, as is generally done to enable backwards binary compatibility, the effective switched capacitance per cycle of the processor, C_{CPU}, will increase due to the implementation of the hardware scheduler. Also, there will be extra capacitive overhead due to branch prediction, operand bypassing, bus arbitration, etc. There will be additional capacitance increase because the N instructions are fetched simultaneously from the cache and may not all be issuable if a branch is present. The capacitance switched for un-issued instructions is amortized over those instructions that are issued, further increasing C_{CPU}.

The energy-efficiency increase can be analytically modeled. Equation 3.1 gives the *ETR* ratio of a superscalar architecture versus a simple scalar processor; a value larger than one indicates that the superscalar design is more energy efficient. The S term is the ratio of the throughputs, and the C_{CPU} terms are from the ratio of the energies, which is proportional to the effective switched capacitance since the architectural comparison is at constant supply voltage. The individual terms represent the contribution of the datapaths, C_{CPU}^{Dx}, the memory sub-system, C_{CPU}^{Mx}, and the dynamic scheduler and other control overhead, C_{CPU}^{Cx}. A 0 suffix denotes the scalar implementation, while a 1 suffix denotes the superscalar implementation. The quantity C_{CPU}^{C0} has been omitted, because it has been observed that the control overhead of the scalar processor is minimal: $C_{CPU}^{C0} \ll C_{CPU}^{D0,M0}$ *[3.5]*.

$$\frac{ETR|_{Scalar}}{ETR|_{Superscalar}} = \frac{S(C_{CPU}^{D0} + C_{CPU}^{M0})}{(C_{CPU}^{C1} + C_{CPU}^{D1} + C_{CPU}^{M1})} \quad \text{(EQ 3.1)}$$

Simulation results show that C_{CPU}^{C1} is significant due to control overhead and that C_{CPU}^{M1} is greater than C_{CPU}^{M0} due to un-issued instructions, thereby negating the *ETR* increase due to S. Since C_{CPU}^{C1} increases quadratically as the number of parallel functional units is increased, the largest improvement in energy efficiency would be expected for moderate amounts of parallelism. In this best case, however, the superscalar architecture yields no improvement in energy efficiency *[3.9]*.

Superpipelined Architectures

These architectures also exploit ILP and offer speedups similar to those found in superscalar architectures *[3.13]*, but their performance is lower because the number of stall cycles increases with the depth of the pipeline due to data dependencies. While these architectures do not need as complex hardware for the dynamic scheduler (C_{CPU}^{Cx} is lower), they do need extra hardware for more complex operand bypassing (C_{CPU}^{Dx} is higher). The net differences in speedup and capacitance give superpipelined architectures an energy efficiency similar to superscalar architectures.

VLIW Architectures

These architectures best exploit ILP by exposing the underlying parallelism of the hardware to the compiler's scheduler, which minimizes the complexity of the hardware. A good compiler is necessary to fully utilize the hardware. One such implementation from Multiflow gives a speedup factor, S, between two and six *[3.17]*.

Because the parallelism is visible, VLIW processors do not require aggressive branch prediction, dynamic schedulers, and complex bus arbitration, so that the energy consumed per operation is roughly the same as for the scalar processor. The main additional energy cost is for the communication network that connects the autonomous functional units that comprise the VLIW processor, and executing the instructions that shuffle data between them. Even with a pessimistic estimate of 50% for the energy per operation increase, the VLIW processor's energy efficiency increases anywhere from 33% to 300%.

Architecture comparison

Superscalar and/or superpipelined architectures are commonly used today because of the increase in performance while maintaining backward machine code compatibility. Unfortunately, utilizing these architectures actually degrades processor energy efficiency. The most energy-efficient processor available today is a simple scalar design with a five-stage pipeline *[3.14]*. While VLIW architectures demonstrate very promising improvement in processor energy-efficiency as defined by the ETR, the required change in the ISA and the increased requirements on the compiler severely complicates their usage.

3.2.3 Modifications for Dynamic Voltage Scaling

While most of the energy efficiency optimizations are relevant whether the Dynamic Voltage Scaling (DVS) described in Section 2.4.4 is used or not, a key modification required at the system level is a DVS-compatible processor bus, which requires the bus to operate across the entire range of dynamically varying operating voltages and performance. Existing processor bus specifications do not support this, so a custom solution was developed, which in turn required custom support as shown in Figure 3.4. In this architecture, only three chips require custom implementation to fully support DVS: the microprocessor, the external RAM, and the I/O interface chip. For the targeted application of portable electronic systems, an important energy consumer is the transfer of data on the processor bus between the microprocessor and the main memory since the I/O bandwidth of typical devices is in the range of 10 kB/sec to 5 MB/sec, which is a small fraction of the available 200 MB/sec peak bandwidth on the processor bus.

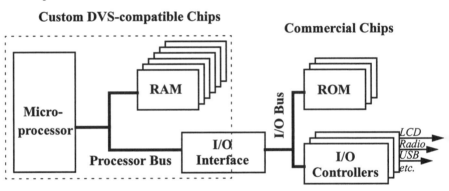

FIGURE 3.4 Prototype System Architecture Incorporating DVS.

The conventional use of a PAL/PLD for I/O control was not made because they use an inordinate amount of power in a typical embedded processor system given the functionality that they provide. Also since commercial PAL/PLD's are not DVS-compatible they were incompatible with the processor. As such, this functionality (memory controller, interrupt controller, etc.) was implemented on-chip within the system-control coprocessor, providing a significant reduction in system energy consumption.

3.2.4 Microarchitecture optimizations

The processor controller typically keeps track of which pipeline stages are being used each cycle. Those pipeline stages not used should have their clock disabled for that cycle. This is particularly important to do in superscalar architectures that typically have only a fraction of the entire processor being utilized. With only a small overhead cost, this technique improves processor energy efficiency by 15-25% (assuming that 40-50% of the processor is disabled 40-50% of the time) *[3.16]*.

To maximize the benefit of clock-gating, null or no-operation (NOP) instructions should be suppressed. In many microarchitectures, NOP instructions are mapped to real instructions. Although NOPs write to a null register, they consume more than half the energy of a normal instruction, as demonstrated by empirical measurements *[3.15]*. Instead, NOPs should be detected by a comparator in the instruction decode stage, and later stages executing on the NOP should be disabled. Similarly, pipeline stalls and/or bubbles should not inject NOP instructions into the pipeline but should instead cause subsequent pipeline stages to be disabled during the appropriate cycle.

Correlation of data is often exploited for energy efficiency in signal processing circuits. While processors do not exhibit the same level of correlation as found in DSP circuits, high amounts of correlation can be found during effective address calculations, which are typically offset from a high-valued stack pointer. In most scalar processors, a single ALU calculates the effective addresses and all integer additions. By partitioning these two types of additions onto separate adders, the signal correlation increases by 16%, decreasing the adder's energy consumed per addition by an approximately equivalent 16%. Total processor ETR can then be improved by 3-7%.

3.2.5 Upper Bounds on Energy Efficiency

The bare essence of a processor is the ability to perform computational operations on data values. This capability, in its simplest form, is performed by the microprocessor's ALU, which is typically a very small fraction of the overall silicon area of the processor. In the prototype system, the fraction is well below 1%. All the surrounding circuit infrastructure merely enables the programmable nature of a general-purpose microprocessor.

Figure 3.5 plots the inverse of energy consumption, MIPS/Watt, for the prototype processor's most elementary component, the adder, and demonstrates how each added level of complexity required for implementing a complete microprocessor increases its energy consumption. Energy consumption is measured at each complexity level, using switch-level circuit simulations, when the processor is executing a common sequence of code. The adder, or similarly the shifter, can operate at 280,000 MIPS/W at 1.2V in our 0.6μm process, and sets a lower bound on energy consumption. Data storage for the adder or shifter's operands necessitates the need for a register file. Including a 30x32b register file increases energy consumption by 2.7x, yielding a more practical upper limit of 105,000 MIPS/W.

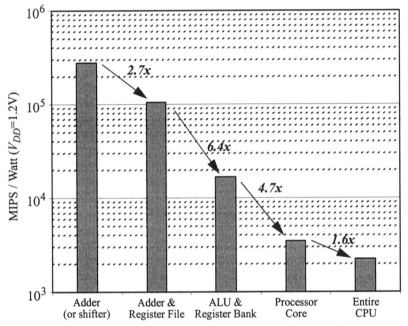

FIGURE 3.5 Energy Consumption for the CPU and its Base Components

The biggest jump in energy consumption, a factor of 6.4x, is attributable to the additional hardware required to build a fully programmable ALU and register bank. This hardware includes latches, muxes, bus drivers, clock drivers, and associated control circuitry. Additionally, the ALU includes a logic unit, a zero-detect unit, and a fast 4-bit shifter, while the register bank includes a 5b→32b register file decoder for each of its two read ports and one write port. The second biggest jump in energy consumption, a factor of 4.7x, is due to the additional hardware of the processor core. This hardware supports instruction fetches, branches, branch prediction, loads, stores, and other ISA-specific functionality. The processor core that was implemented operates at only 3,500 MIPS/W.

Thus, the hardware overhead required to implement a fully-programmable microprocessor core increases energy consumption 30x above that required for just the base adder/shifter and register file circuits. While this overhead will always dominate the total processor energy consumption, this indicates there is significant room for improving a processor's energy efficiency over that of conventional designs. However, what this entails is redesigning the processor core from scratch, starting with the ISA, and progressing down to the microarchitecture, while making design decision designs based not only on improved performance, but reduced energy consumption as well. Measures for both performance and energy consumption are captured when utilizing the *ETR* metric to evaluate the most effective means of improving energy efficiency.

3.2.6 Low-Energy Idle Mode Enhancements

Microprocessors in most single-user system applications, such as notebook computers and PDA's, spend a majority of their time idling as discussed in Section 2.1. Thus, it is essential for the microprocessor to not only reduce the idle energy requirements, but to provide efficient power down modes that can be exploited by an energy-conscious operating system.

The design of the PowerPC 603 processor provides a good demonstration of useful power down modes*[3.16]*. A doze mode stops the processor from fetching instructions, but keeps alive snoop logic for cache coherency and the clock generation and timer circuits, reducing power dissipation by 6.3x for this mode. A nap mode disables the snoop logic, only keeping alive the timer logic, reducing power dissipation another 2.7x. Lastly, there is a sleep mode which only keeps alive the phase-locked loop (PLL) and clock. The power is reduced an additional 17%, while the processor can be up and running at full speed within ten clock cycles, and a cache flush. Further power reduction can be achieved by disabling the PLL in the sleep mode, which reduces power dissipation another 25x (for a total power reduction of 500x), but at the cost of several thousand cycles (up to 200μs) to return to full speed.

It is important to notice how much the PLL, which is found on most microprocessors, limits the reduction of idle energy consumption. Frequently turning off the PLL is not a viable approach due to the large overhead of restarting it.

Microprocessor System Architecture

A significant benefit of DVS is the lack of a high speed PLL. The VCO within the voltage regulation loop is continuously running, but at a very low level of power dissipation, so that the microprocessor can restart with only one cycle of latency. In the prototype system, there was only a single power down mode implemented (Sleep mode), which reduced power dissipation 500x from the peak level, and provided a single-cycle wake-up time.

3.3 Cache System

On-chip cache is essential for providing good system performance and energy efficiency in a general-purpose processor system. An on-chip cache access is simply much faster than an off-chip access to main memory. While cache accesses can typically be achieved at the processor clock rate, external accesses need to traverse through the processor's external interface, the external system bus, and the memory chip's bus interface, all of which increase access latency anywhere from a few processor cycles to tens of cycles [3.18].

A cache access consumes less energy than an access to main memory because the same things that slow the access down -- external bus interfaces and the external bus itself -- also increase its energy consumption. For the prototype system of Chapter 9, a cache access is only 100 pF/access, while an external memory access is 500 pF/access, yielding a 5x reduction in energy consumption per memory access.

However, cache misses require transferring several words into a cache line from main memory, and an equal amount of words need to be transferred from the cache back to main memory when the cache line being replaced is dirty. So what is important to evaluate is the average capacitance/access for a memory access to the on-chip cache, including the effects of cache misses:

$$C_{MA|AVE} = (1 - MR) \cdot C_{CACHE} + MR \cdot LS(C_{CACHE} + C_{EXTMEM}) \cdot (1 + DR) \quad \textbf{(EQ 3.1)}$$

where MR is the fractional cache miss rate, C_{CACHE} is the capacitance of a cache access (100 pF/access), C_{EXTMEM} is the capacitance of an external memory access (500 pF/access), LS is the line size in words, and DR is fractional amount of cache misses that are also dirty. From Equation 3.1, we can calculate what is the maximum miss rate that will ensure $C_{MA|AVE} < C_{EXTMEM}$ so that adding the cache actually improves system energy efficiency. From the prototype system, $LS = 8$ and

Cache System

$DR = 10\%$, such that as long as $MR < 7.7\%$, then $C_{MA|AVE} < C_{EXTMEM}$. This level of miss rate can be achieved with a cache size as small as 2kB *[3.18]*.

So, for almost all practical sizes of an on-chip cache, the inclusion of the cache will not only improve processor performance by reducing the average memory access time, but it will also improve the overall system ETR. The cache used in the prototype design has only a 0.5% miss rate (derived from benchmark program simulations), reducing $C_{MA|AVE}$ to only 125 pF/access, which is still 4x lower than C_{EXTMEM}.

The primary drawback to using a cache is its non-deterministic behavior, which can adversely impact real-time systems. This can be addressed in one of two ways. A software solution entails that any time-critical operation take into account a worst-case latency assuming cache misses. If this is not feasible, a hardware solution consists of either dedicated on-chip memory (e.g. ROM, SRAM) separate from the cache, or a cache which allows specified cache lines to be locked into the cache.

For the PDA-like applications running on the prototype system, a hardware solution to ensure operational latency was not necessary, as the latency issues were sufficiently dealt with in software. Thus, implementing the on-chip cache was essential for improving system energy efficiency, yielding the simplified microprocessor architecture shown in Figure 3.6. The system control block incorporates the glue logic required to seamlessly connect together the custom chips of the prototype system.

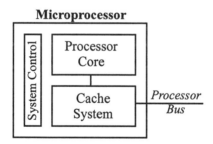

FIGURE 3.6 Prototype Microprocessor Architecture.

3.3.1 Cache size

Increasing the on-chip cache size will always decrease the cache miss-rate, thereby decreasing the number of external accesses and reducing the average memory access time [3.18]. Thus, on-chip cache size is typically maximized, given die-size constraints, for performance considerations.

Maximizing cache size will increase the capacitance/access to the cache. However, this increase can be mitigated in two ways so that capacitance/access does not scale proportionally to cache size. By sub-blocking the cache (Section 3.3.2), the actual memory array size being accessed will remain constant, independent of cache size. Although the interconnect capacitance increases with cache size, a hierarchically buffered bus structure will limit the increase to scale approximately logarithmically. Thus, if buffers are judiciously utilized, the average capacitance/access of a memory access ($C_{MA|AVE}$) will continue to decrease with cache size past 256kB as demonstrated in Figure 3.7. If, however, the interconnect capacitance scales proportionally with cache size, then there is an optimal cache size, beyond which the average capacitance/access increases.

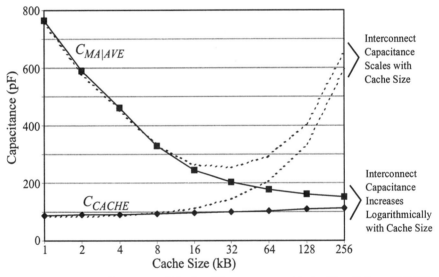

FIGURE 3.7 Average and Cache pF/access versus Cache Size (0.6μm CMOS).

In summary, maximizing the cache size will generally maximize system energy efficiency by providing the highest performance and the lowest energy/access. For the prototype system, a cache size of 16kB was chosen, and was strictly limited by die-size constraints. In Figure 3.7, $C_{MA|AVE}$ was calculated based upon the cache miss rates from [3.18]. For the prototype system described in Chapter 6, benchmark simulation reported a cache-miss rate of 0.5% for a 16kB cache. This reduces $C_{MA|AVE}$ from the 250 pF/access shown in Figure 3.7 (based on a 2.0% miss rate) down to 125 pF/access

3.3.2 Sub-blocking

Most large SRAM's are not composed of a single memory array, due to excessive delay on the word and bitlines, but are composed of several smaller arrays or sub-blocks in order to improve memory access time [3.19]. Enabling only the desired block, rather than the entire memory array also reduces energy consumption. [3.20][3.21].

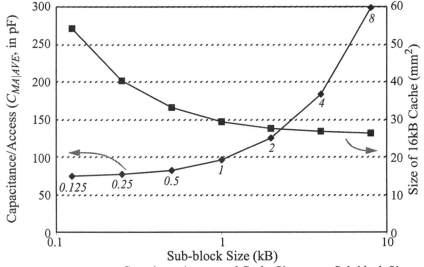

FIGURE 3.8 Capacitance/access and Cache Size versus Sub-block Size.

Reducing the size of the basic memory sub-block will reduce the energy consumed per access as well as speed up the access time. However, circuit overhead consisting of control signal generation, sense-amps, and output buffers sets the lower limit on feasible sub-block size. Figure 3.8 plots both capacitance/access and area of a 16kB cache as a function of sub-block size in our 0.6μm process. For large sub-block sizes, the overall area asymptotically approaches a fixed value as the fractional contribution

Microprocessor System Architecture

of the circuit overhead goes to zero, while the capacitance/access scales up with sub-block size. For small sub-block sizes, the capacitance/access asymptotically approaches the fixed cost due to the circuit overhead, while the overall cache size begins to exponentially increase. For sub-block sizes less than 0.125kB, the capacitance/access will actually begin to increase due to the I/O capacitance loading from the additional sub-blocks. For the prototype system, a size of 1kB was chosen, which adequately balanced the capacitance/access versus the overall cache size. At this design point, the cache contributes 30% of the total processor energy consumption, and 50% of the silicon area.

3.3.3 Tag Memory Architecture

In addition to the data memory, which hold the contents of external memory locations, an integral part of a cache is the smaller tag memory, which is used to map cache locations to the global memory space. Multiple sequential data words are generally organized into a cache line, which is then mapped as a single cache location into the global memory space. This amortizes the cost of the tag memory required over a larger data memory size*[3.18]*.

The width of the tag memory is:

$$\text{Tag Width} = \log_2 \left(\frac{\text{Address Space}}{\text{Cache Size}} \cdot \text{Set-Associativity} \right) \quad \text{(EQ 3.2)}$$

where the *Address Space* for the ARM8 architecture is 32 bits. Thus, the *Tag Width* is a minimum of 18 bits for a direct-mapped 16kB cache, and the number of bits will increase with the $\log_2$ of the set associativity. The number of tags required is:

$$\text{Number of Tags} = \frac{\text{Cache Size}}{\text{Cache Line Size}} \quad \text{(EQ 3.3)}$$

For the 16kB cache and 32B cache line size used in the prototype system, there are 512 tags, such that the total tag memory is on the order of 512 x 20b, or 1.28kB.

If the tags are stored in a single memory array, the energy consumed accessing a tag would be more than the energy consumed accessing the actual data word itself, since the cache is sub-blocked in 1kB arrays. However, the tag memory can similarly be sub-blocked. The same four address bits which are decoded to selectively activate the data memory sub-blocks can be used to selectively activate the tag memory sub-blocks. Then, only a 32 x 20b tag memory array is activated per cache access, significantly reducing the energy consumption per tag memory access. The increase in the memory critical path due to the gate delays added by the four-bit decoder is offset by the reduced access time of a smaller tag memory array.

Design Approaches

During an access to the cache, the requested address is compared against the tag(s) where the address may potentially reside in the cache. If there is a match between the address and a tag, then the contents are currently stored in the cache, and the contents are returned to the processor core. If there is not a match, then that address location needs to be fetched from the external memory system. For an N-way set-associative cache, N tags must be compared against the requested address because the requested address may reside in any of N cache line locations. There are three general approaches to implementing the tag compare:

Serial RAM Access: The tag memory is accessed first. If any of the N tags matches the requested address, then an access to the data memory is initiated. The benefit of this approach is that the data memory is only accessed for a cache hit. However, the disadvantage is that the serial RAM accesses may dominate the critical paths of the processor.

Parallel RAM Access: The tag and data memories are accessed in parallel. Thus, if any of the N tags match the requested address, the desired data word has already been read from the data memory and is available. However, writes to the data memory still must be done sequentially since the write cannot be committed to cache memory until the tag has been validated. The benefit of this approach is a fast read access, which are the dominate type of cache accesses (> 80%). The disadvantages are that cache writes must still be done serially, potentially requiring an additional cycle to complete, and extra energy is consumed for data memory reads on cache misses.

Serial CAM/RAM Access: The tag memory is implemented as a content-addressable memory (CAM). The tag CAM is first accessed, and if a match is found, then the data RAM is accessed. The benefit of this approach is that the CAM's match signals can be used directly as word line enables for the data RAM. This eliminates the need for an address decoder for the data memory, significantly speeding it up so that this approach does not dominate the critical paths of the processor. In addition, the data memory array is only accessed upon a cache hit. The disadvantage is that CAMs generally consume significantly more energy than their RAM equivalents.

Microprocessor System Architecture

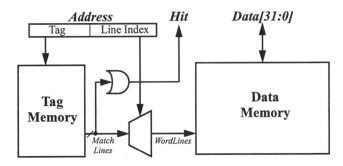

FIGURE 3.9 Optimized CAM/RAM Cache Memory Architecture.

Optimized Architecture

A CAM array was optimized and fabricated, which only consumes twice the energy of an equivalently-sized, 32 x 20b SRAM memory array. This removed the key disadvantage to the Serial CAM/RAM Access approach, and therefore made it the optimal design approach to select.

By utilizing a CAM for the tag memory, 32 tags can be compared simultaneously for the same energy consumed as when comparing one tag. Thus, if the cache is 2-way set-associative, the energy consumed by the Serial CAM/RAM Access approach is the same as for the Serial RAM Access approach, because in the latter, two RAM accesses must be made to read two tags. With the CAM, the comparison between the tag and the requested address is implicit, thereby removing the need for external comparators as in the Serial RAM Access approach. Also, the Serial CAM/RAM Access approach eliminates the need for an address decoder for the data memory. Thus, for 2-way and higher set-associativity, the Serial CAM/RAM Access approach has the lowest energy/access.

With respect to access time, the Serial CAM/RAM Access approach is only slightly slower than the Parallel RAM Access approach. With the CAM, the removal of the data memory address decoder and external comparators reduces the cycle time to within 10% of the Parallel RAM approach. Therefore, for 2-way and higher set-associativity, the Serial CAM/RAM Access approach is the most energy-efficient solution for the implementation of the tag memory. Since some level of associativity was required for the prototype design, as will be explained in Section below, the Serial CAM/RAM Access approach is the optimal solution.

The basic architecture is shown in Figure 3.9. The upper bits of the address, which constitute the tag, are sent to the tag CAM array. If the tag is present in the CAM, one of the match lines will go high, and indicate a cache hit. The line index bits of the address are used to demultiplex the match line to the appropriate word line of the data memory array.

3.3.4 Associativity & Cache Line Size

For the ARM8 processor core used in the prototype system, the memory interface was designed and optimized for a unified cache *[3.22]*. Given the constraint that the cache must be unified, a direct-mapped cache is not very desirable. When instruction and data addresses point to the same cache line, which may frequently happen, the cache will continue to alternately swap out of cache memory the conflicting address locations until the conflict is removed, needlessly spending many cycles transferring the cache lines. Thus, at least 2-way set associativity is desirable to prevent this conflict.

The requirement for at least 2-way set associativity dictated that the optimal tag memory architecture is the Serial CAM/RAM Access approach. With this architecture, the set associativity and cache line size for a fixed 1kB sub-block are related:

$$1\text{kB} = \text{Set-Associativity} \cdot \text{Line Size} \qquad \textbf{(EQ 3.4)}$$

Simulations were then used to find the optimal cache line size and set-associativity with respect to energy consumption, as shown in Figure 3.10.

To measure the true impact on system energy consumption, the capacitance/access was measured for not only the cache, which is broken down by tag and data memory, but also for the external memory system. As the line size increases, the capacitance/access contributed by the tag array decreases due to a smaller tag memory array, and because the tag memory array does not need to be accessed for sequential cache accesses to the same cache line.

However, the capacitance/access contributed by the external memory increases because more words are being fetched per tag, not all of which may be used by the processor core. The contribution from the data memory array is relatively constant because it is dominated by processor core cache accesses, which is independent of cache line size, rather than cache-line reloads which are much more infrequent. Due to the low miss rate of the 16kB cache, the impact on performance is negligible for

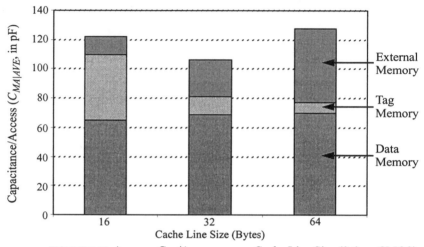

FIGURE 3.10 Average Cap/Access versus Cache Line Size (0.6μm CMOS).

these cache-line sizes and levels of set associativity. Thus, the optimal cache-line size for energy efficiency is 32B, which gives the cache 32-way set associativity.

Simulation demonstrated that a 4-way set-associative cache would provide an insignificantly higher miss rate than a 32-way set-associative cache, but the high degree of associativity is set by a combination of the Serial CAM/RAM Access tag memory architecture, the 1kB cache sub-block size, and the optimal cache line size of 32B.

The $C_{MA|AVE}$ reported in Figure 3.10 for the 32B line size used in the prototype system is 106 pF/access, which is below the $C_{MA|AVE}$ of 125 pF/access as reported in Section 3.3.1. This reduction of 19 pF/access occurs because this simulation better models the tag memory, which is not activated for sequential memory accesses to the same cache line.

3.3.5 Cache Policies

Simulation of the processor system running the benchmark programs described in Section 10.4 was utilized to evaluate the most energy-efficient choice for the following cache policies.

Write Policy

The write policy dictates what happens upon a write for a cache hit. A write-through policy dictates that every write is also transferred to the external memory to maintain continuous memory consistency between the cache and the external memory. A write-back policy only writes the data words back to main memory when a cache line that had previously been written to is removed from the cache to be replaced by another cache line.

These two policies demonstrate negligible impact on system performance, but do show a difference in the total system energy consumption as shown in Figure 3.11. For all three benchmarks, a write-through policy yielded consistently higher total system energy consumption, due to an increase in external processor bus activity. With the write-back policy, a cache line can be written to multiple times before being sent to main memory, thereby decreasing the external memory traffic. Thus, the more energy-efficient write-back policy was selected for the prototype system.

Write Miss Policy

The write miss policy dictates what happens upon a write for a cache miss. A write-allocate policy will allocate space in the cache for the address being written. The cache line is first placed into the cache, and then the write can be completed. A read-allocate policy will not allocate space, but send the data word directly to main memory.

Simulations demonstrate negligible impact on both performance and energy consumption. Thus, the policy chosen was read-allocate, which simplified the design of the cache controller due to the complexity of the cache line allocation process, as described in Section 6.2.2. External memory can support both byte and word writes, so the only added complexity was to support half-word writes, as specified by the ARM8 ISA, which are broken into two separate bytes before they are written out to main memory.

Replacement Policy

The replacement policy dictates which cache line gets replaced during the cache line allocation of a read miss. There are several choices of replacement policy. One common policy is least-recently-used (LRU) replacement, which swaps out the cache line which has not been accessed for the longest time. Another policy is round-robin replacement, which cycles through the potential cache lines. A third common policy is random replacement.

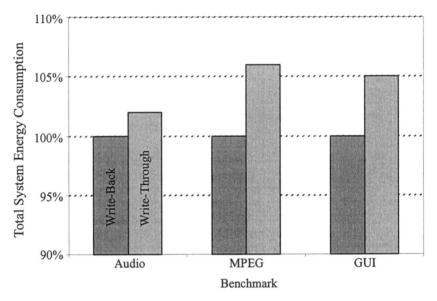

FIGURE 3.11 Energy Consumption vs. Write Policy for Benchmark Programs.

For moderately sized caches (< 64kB) and for much lower degrees of set associativity (e.g. 2-way or 4-way), the LRU replacement policy would provide the most energy-efficient policy by minimizing the cache miss rate[3.18]. However, for the high degree of set associativity (32-way) for the prototype system, simulations demonstrate negligible impact on both performance and energy consumption for all three policy variants. Thus, the simplest policy for implementation was again selected, which is round-robin. This can be implemented with 1 latch per cache line, which keeps track of the last replaced line within each cache sub-block, and advanced one cache line upon a replacement.

Level-0 Cache

A level-0 (L0) cache is essentially a small buffer between the primary (L1) cache and the processor core. These are generally used when the primary cache cannot complete a memory access within one cycle, due to either a very fast processor clock speed or a very large primary cache memory [3.18]. In addition, L0 caches have been demonstrated to improve the energy efficiency of cache systems[3.20][3.21]

An L0 cache contains a data and tag memory, similar to the primary cache. With an L0 cache, a memory access first checks the L0 tag memory to see if it contains the desired memory location. If so, then the L0 cache returns the desired word and the primary cache does not need to be activated. If not, once the desired memory contents have been located in the memory hierarchy, the cache line is placed into the L0 cache. If the L0 cache size is only one cache line, then the L0 can be implemented with little impact on the cache system's performance since only a single tag compare is added into the critical path. However, larger L0 cache sizes, which need to be fully-associative, require additional hardware complexity that will increase the capacitance/access of an L0 cache hit, and may bloat the critical path and force a L0 cache miss to extend over an additional clock cycle.

With the slight modification to the CAM/RAM architecture highlighted in Figure 3.12, it can support implicit buffers, which are functionally equivalent to an L0 cache, by latching the match lines and storing the location of the previously matched tag. If a cache access has the same tag as the previous cache access, then the tag memory does not need to be enabled, and the saved match line already points to the correct word line of the data memory.

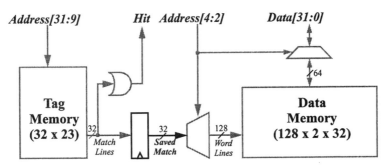

FIGURE 3.12 Implementing an Implicit Buffer in the CAM/RAM Architecture.

While each cache 1kB sub-block has this implicit buffer, to keep all of these buffers active would require a 16-entry hash table of the last tag access to each sub-block. The complexity then becomes similar to a larger L0 cache. To simplify the hardware, only one implicit buffer can be kept active at a time, requiring the storage of only the tag of the previous cache access. This is functionally equivalent to a one cache line L0 cache.

Figure 3.13 demonstrates the improvement in energy efficiency of the L0 cache and implicit buffer by plotting the capacitance/access of a cache hit as a function of their hit rates, and comparing them to a baseline implementation which accesses the primary cache tag memory on each cache access (i.e. no L0 cache). Two sizes of L0 caches are compared, as well as the implicit buffer approach. A high-level simulation is used to estimate the capacitance/access by combining the capacitance/access of the individual blocks with their activity factors. For low L0 hit rates, the L0 cache approach is much less energy-efficient than even the nominal case because many words are being fetched from the primary cache and placed into the L0 cache, which never get used by the processor core. Only for high hit rates does the L0 cache become more energy efficient. The implicit buffer is more energy efficient across the broadest range of hit rates, and is always more energy-efficient than the nominal case.

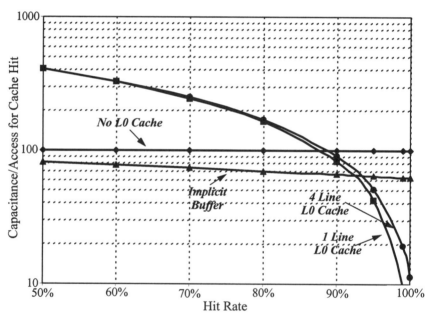

FIGURE 3.13 Capacitance/Access of a Cache Hit for L0 Architectures

For the prototype system, the implicit buffer approach was chosen. For a one-line (32B) L0 cache, the hit rate will be on the order of 80% *[3.3]*, at which value the implicit buffer has 42% of the capacitance/access of a one-line L0 cache. A four-line L0 cache can achieve much higher hit rates, in excess of 93.5%, at which point it

Cache System

becomes more energy-efficient than the implicit buffer approach. However, the added design complexity of the four-line L0 cache outweighed the energy-efficiency improvement. The implicit buffer approach provides a 30% improvement in energy efficiency with minimal additional design complexity.

Improvement with a Write Buffer

Write stalls occur when the processor core has to be halted while it is waiting for an external write to complete. These occur for cache misses in which the cache line being discarded has been modified in the cache and needs to be updated in external memory. In addition, this occurs for direct writes to external memory which bypass the cache. These typically occur for writes to the I/O memory space, which are quite common in embedded processor systems. The use of a write buffer can eliminate these write stalls by holding the stores and cache line writes and sending them to external memory during free external bus cycles, thereby eliminating the write penalty that is otherwise present (20 core cycles per cache line, 6 core cycles per single-word write).

The basic implementation is shown in Figure 3.14. The write buffer collects external memory writes from the cache memory and processor core at the processor clock rate. When the external bus interface is free, the write buffer then completes the writes at the external bus clock rate.

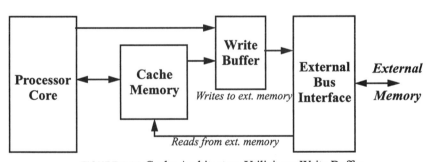

FIGURE 3.14 Cache Architecture Utilizing a Write Buffer.

For cache line writes, the system performance improvement is minimal due to the low miss rate of the 16kB cache. The number of processor memory accesses in which a cache line needs to be written out is 0.1%, thereby reducing the processor cycles per instruction (CPI) by only 2%. However, the percentage of instructions that are writes to I/O space is 1% for the benchmark programs, which translates into a reduction of

15% in CPI, which is significant. Since the write buffer is used for only a small fraction of the time, its energy consumption is negligible (< 0.2%). Thus the inclusion of a write buffer improves the system ETR by at least 15%.

3.3.6 Interfacing to an External Bus

The unified cache simplifies the external bus interface as compared to split instruction/data caches, in which snoop hardware is required to maintain memory coherency between the split caches *[3.18]*. With the addition of the write buffer, the cache system busses can be connected directly to the bus interface with no adverse affects on performance.

Bufferable writes are placed into the write buffer at the internal CPU clock rate, and while the processor continues to operate, the bus interface writes the words in the write buffer out to main memory. For reads that must go out to main memory, the processor is halted until the desired word is available in the bus interface, so no performance is lost.

3.3.7 Advantages and constraints of the ARM8 memory interface

The biggest constraint of the ARM8 memory interface is that it is designed and optimized for a unified cache, which is generally less energy-efficient than split instruction/data caches. Since a unified cache requires some level of associativity, a CAM-RAM tag memory architecture was implemented, and provided 32-way set-associativity. A unified cache has one key advantage, which is that the hardware required to maintain coherency in split caches is eliminated. The back-side of the cache can communicate directly with the bus interface and external memory.

The ARM8 memory interface contains a complex request-acknowledge handshake protocol which can be utilized to improve the performance and energy-efficiency of the cache subsystem *[3.22]*. Encoded in the request control signal is the type of request (load, store, instruction fetch), what size it is (byte, halfword, or word), whether the request is sequential to the last memory request, whether there are more words to follow (as part of a load/store multiple instruction), and whether the instruction fetch is speculative or not. Encoded in the acknowledge signal is whether the request completed or aborted, and how many words were successfully returned on a load/fetch request.

Cache System

In addition, the cache system can send two additional signals back to the ARM8 core indicating whether the instruction and data buffers contain valid data from a previous cache load/fetch request. In the prototype system, implicit buffers were implemented inside the cache blocks themselves, as described in Section 6.2.2. If the cache indicates to the ARM8 core that the end of the buffer has not yet been reached, then the core does not need to place the address on the internal Address bus for sequential loads/fetches, saving significant energy by not needlessly driving the bus.

3.3.8 An ARM8-optimized cache system

In addition to tuning the general properties (e.g. size, associativity) and policies (e.g. write, replacement) of the cache system to the ARM8 memory interface, further architectural design choices were implemented to take full advantage of this complex interface.

Double Reads

The drawback of a unified cache is that 25% of the instructions are data transfer instructions, as measured from benchmark simulation. Thus the average number of memory word accesses is 1.25 words per instruction. A standard memory bus can only transfer one word per cycle, which would force the processor core to stall on a data transfer instruction 25% of the time. To prevent the memory bus from being a performance bottleneck, the ARM8 interface allows for two words per cycle to be retrieved from the cache system.

The data RAM in the cache is organized into two columns of 32 bits each, which is multiplexed depending upon the address modulo 2. By only allowing double reads to occur for even word addresses, retrieving the second word simply entails switching the multiplexer and enabling the sense amp. Neither the CAM nor the word-line driver is reactivated to read the second word. This strategy allows two words to be returned per cycle, with the energy cost of retrieving the second word minimized to be less than 40% of a standard cache read. The penalty for this strategy is that odd-word addresses cannot return two words. However, once the prefetch unit is even-word aligned, it remains even-word aligned until the next jump to an odd-word address, so that the impact on performance is negligible.

Sequential Reads

The memory interface encodes whether the address of a fetch/load request is sequential to the previous fetch/load request's address. Benchmark simulation found that only 8% of data accesses are sequential, while 70% of instruction fetches were sequential. Since only 20% of all instructions are loads, the net energy savings of an L0 cache for loads is only 0.25%. However, the energy savings of an L0 cache for instruction fetches is 10%. Thus, the implicit buffer for the L0 cache (Section) was only implemented for instruction fetches, and not data loads, since only the former yields a significant reduction in overall processor energy consumption.

Using the 8-word implicit buffer simplifies the implementation of the virtual L0 cache within cache memory array. Only four bits of state are required to encode which of the sixteen 1kB blocks contains the current instruction buffer location. When the core requests a sequential instruction fetch and the address is at the beginning of a new cache line, the cache controller suppresses the CAM tag array, and the next word in the implicit buffer is read from the data memory array.

Load/Store Multiple Registers

The ARM8 memory interface also encodes whether there are more loads/stores to follow sequentially, as part of a load/store multiple-register instruction (LDM/STM). If the load/store is to the cache, these operations proceed at the core clock rate, and cannot be further optimized.

However, if an STM takes a cache miss, then this encoded information allows the cache controller to packetize multiple words per address, aligned on cache-line boundaries, and place the address(es) and data words in either the write buffer, or send them directly to the external bus interface. This increases the data bandwidth on the external bus and decreases its energy consumption, compared to single-word stores, which require an address to be transmitted on the external processor bus for each data word.

If an LDM takes a cache miss, there is high probability that the missing cache line will be loaded into the cache. Thus, optimizing LDM instructions was not necessary, since the required cache lines will be loaded, and then the LDM will be serviced from the cache at the processor clock rate.

3.4 System Coprocessor

The primary role of the system coprocessor is to configure global processor settings, interface to the voltage converter chip, and maintain the system control state. The coprocessor has very low performance requirements, and consumes very little energy, and can be considered negligible. However, the energy consumption does become critical while the processor is in the sleep mode, and in part, determines the total system sleep-mode power dissipation. Thus, the coprocessor was carefully designed to minimize its standby power dissipation.

3.4.1 Architecture

The ARM8 processor core provides a dedicated coprocessor interface, through which coprocessor instructions are passed to the coprocessor unit and are executed in a parallel 3-stage pipeline. Logically, the coprocessor looks like a large register file which can be read from, and written to, by coprocessor data-transfer instructions (MCR, MRC). However, the registers themselves are very heterogeneous, and cannot be implemented as a standard register file. Some registers are read-only counters, others have hard-coded values, while others are completely virtual in that a write to them initiates some action by the coprocessor.

Thus, the coprocessor was implemented by connecting the registers with a shared input and output bus architecture. The heterogeneous register bitslices were pitched-matched to provide compact layout.

3.4.2 Providing an integrated idle mode

The ARM8 core does not provide a processor halt instruction. To implement this instruction in the prototype processor, a coprocessor write instruction was used to implement this feature. Upon a write to this register, the global clock signal is halted, stopping processor operation. The processor can be restarted via an external interrupt, or an internal timer interrupt.

Since all processor state is maintained during sleep mode, the operating system can seamlessly enter and exit sleep mode without disturbing the state of the currently executing software thread.

Microprocessor System Architecture

3.5 Summary

Energy-efficient architectural optimizations at both the system level and within the cache subsystem significantly improved the overall processor system's energy efficiency, even while limited by the fixed architecture of the ARM8 processor core. A future, energy-optimized processor core may yield even further gains in overall system energy efficiency.

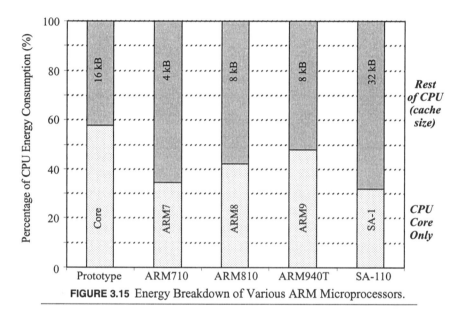

FIGURE 3.15 Energy Breakdown of Various ARM Microprocessors.

In the prototype processor described in detail in Chapter 6, the majority of the architectural optimizations occurred within the cache subsystem and the peripheral circuitry around the processor core. Figure 3.15 plots the fraction of energy consumed by the core, and the remainder of the processor for the prototype chip, and four other implementations of the ARM architecture. The ARM8 core used in the prototype is the same as the processor core in the ARM810, and very similar to the core in the ARM940T, and SA-110. What is significantly different between these chips is the non-core component of the processor, whose energy consumption is dominated by the cache. What is demonstrated in this figure is the improvement in energy consumption of the cache subsystem, which consumes only 42% of the energy in the prototype processor. In the other processor chips, the fraction ranges from 52% to as much as 70% in the SA-110. This shows that the energy-efficient architectural design methodology presented in this chapter can provide significant reduction in energy consumption and because of the similarity of this processor family with others, these techniques will yield similar results in other types of embedded processors.

In the case of the ARM8 processor, whose core is logically equivalent to the prototype processor, the energy of the cache subsystem has been reduced by 30% in relative terms while providing twice as large a cache. In absolute terms, the energy reduction is 61%, or more than a 2x reduction.

References

[3.1] J. Bunda, et. al., "16-Bit vs. 32-Bit Instructions for Pipelined Architectures", *Proceedings of the 20th International Symposium on Computer Architecture*, May 1993, pp. 237-46.

[3.2] Advanced RISC Machines, Ltd., *Introduction to Thumb*, Developer Technical Document, Mar. 1995.

[3.3] J. Bunda, W.C. Athas, and D. Fussell, "Evaluating Power Implications of CMOS Microprocessor Design Decisions", *Proceedings of the 1994 International Workshop on Low-Power Design*, Napa Valley, CA, April 1994.

[3.4] P. Freet, "The SH Microprocessor: 16-Bit Fixed Length Instruction Set Provides Better Power and Die Size", *Proceedings of the Thirty-Ninth IEEE Computer Society International Conference*, Mar. 1994, pp. 486-8.

[3.5] T. Burd, B. Peters, *A Power Analysis of a Microprocessor: A Study of an Implementation of the MIPS 3000 Architecture*, ERL Technical Report, University of California, Berkeley, 1994.

[3.6] J. Montanaro, et. al., "A 160MHz 32b 0.5W CMOS RISC Microprocessor", *Proceedings of the Thirty-Ninth IEEE International Solid-State Circuits Conference - Slide Supplement*, Feb. 1996, pp. 170-1.

[3.7] J. Bunda, *Instruction-Processing Optimization Techniques for VLSI Microprocessors*, Ph.D. Thesis, The University of Texas at Austin, 1993.

[3.8] A. Chandrakasan, S. Sheng, and R.W. Brodersen, "Low-Power CMOS Digital Design", *IEEE Journal of Solid State Circuits*, Apr. 1992, pp. 473-84.

[3.9] R. Gonzalez and M. Horowitz, "Energy Dissipation in General Purpose Processors", *Proceedings of the IEEE Symposium on Low Power Electronics*, Oct. 1995, pp. 12-3.

[3.10] M. Johnson, *Superscalar Microprocessor Design*, Englewood, NJ: Prentice Hall, 1990.

[3.11] D. Wall, *Limits of Instruction-Level Parallelism*, DEC WRL Research Report 93/6, Nov. 1993.

[3.12] M. Smith, M. Johnson, and M. Horowitz, "Limits on Multiple Issue Instruction", *Proceedings of the Third International Conference on Architectural Support for Programming Languages and Operating Systems*, Apr. 1989, pp 290-302.

[3.13] N. Jouppi and D. Wall, "Available Instruction-Level Parallelism for Superscalar and Superpipelined Machines", *Proceedings of the Third International Conference on Architectural Support for Programming Languages and Operating Systems*, Apr. 1989, pp 272-82.

[3.14] J. Montanaro, et. al., "A 160-MHz 32-b 0.5-W CMOS RISC Microprocessor", *IEEE Journal of Solid State Circuits*, Vol. 31, No. 11, Nov. 1996, pp. 1703-14.

[3.15] V. Tiwari, et. al., "Instruction Level Power Analysis and Optimization of Software", *Journal of VLSI Signal Processing*, Vol. 13, Nos. 2/3, Aug/Sep 1996, pp. 223-238.

[3.16] S. Gary, et. al., "The PowerPC 603 Microprocessor: A Low-Power Design for Portable Applications", *Proceedings of the Thirty-Ninth IEEE Computer Society International Conference*, Mar. 1994, pp. 307-15.

[3.17] P. Lowney, et. al., "The Multiflow Trace Scheduling Compiler", *The Journal of Supercomputing*, Vol. 7, Boston: Kluwer Academic Publishers, 1993, pp. 51-142.

[3.18] J. Hennessy, D. Patterson, *Computer Architecture: A Quantitative Approach*, Morgan Kaufmann, San Francisco, 1995.

[3.19] J. Rabaey, *Digital Integrated Circuits, A Design Perspective*, Prentice Hall, Upper Saddle River, NJ, 1996.

[3.20] J. Bunda, W.C. Athas, and D. Fussell, "Evaluating Power Implications of CMOS Microprocessor Design Decisions", *Proceedings of the 1994 International Workshop on Low-Power Design*, Napa Valley, CA, April 1994.

[3.21] C. Su, A. Despain, "Cache Designs for Energy Efficiency", *Proceedings of the Twenty-Eighth Hawaii International Conference on System Sciences*, Jan. 1995, pp. 306-315.

[3.22] Advanced RISC Machines, Ltd., *ARM 8 Data Sheet*, Document Number ARM-DDI-0100A-I, Feb. 1996.

CHAPTER 4 *Circuit Design Methodology*

The key to energy-efficient circuit implementation, much like architecture and system design, is to focus on energy consumption throughout the entire process, rather than addressing it only as the design nears completion. There are a number of simple rules that will yield an energy-efficient circuit implementations that will be described in this Chapter. Also, if a set of simple rules are followed most CMOS circuits can be made to be robust against the voltage variations of DVS which will also be presented. To demonstrate these principles the design of several complex blocks such as the arithmetic and memory circuits will be described in detail. Because of the importance of bus transceivers, a discussion of an ultra-low energy design will also be given.

4.1 General Energy-Efficient Circuit Design

This section will describe a set of circuit design techniques which apply equally well to any digital CMOS integrated circuit. Many of these design techniques were first developed for low-power, custom DSP ASIC's *[4.1]*, and have been applied here to a general-purpose processor system.

In an ideal digital system, all signal paths through the circuits have equal delays, but in any practical system, this is not the case. Typical there is a small fraction of signal paths that determine the achievable cycle time, the critical paths, and in those paths, increased energy consumption may be warranted to decrease circuit delay to meet a target cycle time. All other paths should consume as little energy as possible. For some paths, once their delay is increased to the cycle time, making them a critical path, no further energy reduction can take place. Other paths, typically with very small logic depth, have considerable slack and should be optimized solely for energy.

To make these optimizations, the circuit schematics must be analyzed and modified before being committed to layout, after which, changes become much more time intensive. Some paths reside entirely within a block, such that they can be optimized solely within the block design. Other paths cross over multiple blocks, requiring a complete schematic design of all dependent blocks for a truly optimal design, rather than simply relying on predetermined setup and hold delays based upon a behavioral model, and optimizing each block individually. The design methodology is described in much more detail in Chapter 5.

4.1.1 Logic Style

There are a variety of logic styles to choose from, such as static CMOS, CPL, Domino, NORA, C^2MOS, CVSL, etc., which vary widely in their delay and energy consumption *[4.2]*. The optimal logic style cannot be found by merely selecting the one that has the smallest total capacitance. They must be compared by analyzing their effective switched capacitance per cycle, which factors in signal transition frequencies. The outputs of static CMOS and CPL only transition upon an input transition, while dynamic logic styles (Domino, NORA, etc.) incur output transitions both upon input transitions, and during the precharge phase of every clock cycle. The clock nodes in dynamic circuits have an energy-consuming transition every cycle, too. So, while dynamic logic styles tend to be faster, they can result in increased energy consumption.

For simple cells (e.g. AND, OR, AOI, etc.), the optimal logic style with respect to energy efficiency is generally static CMOS *[4.1]*, while for more complex cells, there is no single, optimal logic style so that it is important to investigate a variety of logic styles for the particular function and requirements. Additionally, DVS places further restrictions on logic design.

DVS Compatible Logic Design

While static CMOS is fully compatible with DVS, dynamic logic styles require some modification to ensure proper operation. Fortunately, these modifications have little impact on circuit performance and energy consumption. For buffered dynamic logic styles, which are predominantly used in the prototype design, a small bleeder PMOS device is added to maintain state on the precharged node when the inputs to the pull-down network are not actively pulling the node down, as shown in Figure 4.1. This device can have minimum width and non-minimum length, as very little current is required to maintain state.

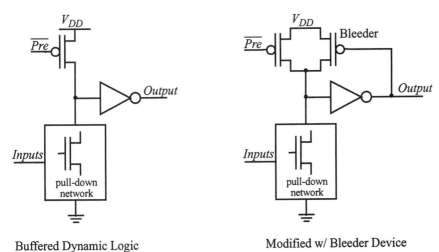

Buffered Dynamic Logic
(e.g. Domino)

Modified w/ Bleeder Device

FIGURE 4.1 Energy efficient bleeder circuit for dynamic Logic.

In the prototype system, all synthesized logic utilizes static CMOS logic. Dynamic logic was only selectively utilized in custom-designed blocks, such as wide-NOR gates for zero-detection operations, wide-AND gates for decoding, and shifter gates, where the effective switched capacitance reduction more than compensated for the increased switching activity.

ALU Design Example

Since the delay of CMOS circuits scales well over voltage, the initial circuits were designed at 3.3V which set the target cycle time at 10ns. There were two critical data paths in the ALU, both of which had only a half-cycle, or 5ns, to complete.

One path consists of a simple shift (0, 1, 2, or 3 bits only), a selective inversion, and a 32 x 32 adder. The critical element in this path was the adder. The shifter was implemented with a four-way mux, and the selective inversion with an OR gate for a total

Circuit Design Methodology

of three gate delays. This allowed approximately 3.5ns for the addition to complete, taking into account latch setup and hold time requirements. Various adders were analyzed, including ripple, carry-select, and Brent-Kung adders *[4.2]*. The latter was selected because it could meet the targeted delay in the minimal energy consumption. Other adders, such as the ripple-carry, had lower energy consumption, but were removed from consideration because they could not meet the delay target.

The other path consists of a fully-programmable 32-bit shift, and a logic operation unit. The logic operation unit (AND, OR, XOR) maps to a single combinational logic gate, with an additional gate required for buffering. The shifter had approximately 3.5ns to complete its operation, as well. The natural implementation of the shifter would be a barrel shifter, which is the most compact. However, for DVS compatibility, the usual NMOS pass gates must be replaced with CMOS pass gates. This causes the 32-bit barrel shifter to consume a large amount of energy due to the large CMOS pass gates required to keep delay through the shifter minimized. Instead, a logarithmic shifter was utilized to reduce energy consumption, and was tuned to meet the target cycle time.

4.1.2 Transistor Size

Traditional design methodologies utilize cell libraries with transistor sizes larger than necessary. A typical "1x" output driver size uses transistor widths much larger than the minimum size. This is due to the desire to provide maximum drive capability under a wide variety of load conditions. While this increases gate-area density and simplifies the cell libraries required, it is not energy-efficient.

A more energy-efficient solution is to set the base "1x" driver size to be minimum size. For our MOSIS 0.6μm process, the minimum NMOS width is 1.2μm. To equalize rise and fall times, and minimize gate delay, the "1x" PMOS width is 2.4μm due to the lower mobility of PMOS devices. A simple, energy-efficient, transistor-sizing methodology is to initially set the size of all transistors so that short-circuit current is minimized, as will be described in Section , below. Transistors in the critical paths are then increased in size to decrease delay, as necessary, and all remaining transistors with small fan-out are reduced in size while not violating constraints for minimizing short-circuit current.

Conventional belief is that as process technology improves, interconnect capacitance will dominate the total capacitance on a node, making transistor-size dependent capacitance (gate oxide and diffusion capacitance) insignificant. Thus, the optimal transistor size is much larger than minimum size, since performance will increase while having a negligible impact on energy consumption. However, this is not true, since while interconnect capacitance will dominate the global nets, for local nets, transistor parasitic capacitance will continue to be significant, and remain critical to minimize whenever possible*[4.3] [4.4]*.

Minimizing Short-Circuit Current

By bounding the ratio of input to output rise/fall times between gates, short-circuit current energy consumption can be minimized. If the ratio is kept to less than two, the upper limit of additional energy consumption is 12% at $V_{DD} = 3.3V$. This is achieved by increasing the size of devices as necessary when driving large loads. This constraint will be defined as:

$$MIORFT = \text{Maximum Input-to-Output Rise/Fall Time} = 2 \quad \text{(EQ 4.1)}$$

The simplest gate construct consists of minimum-size, back-to-back inverters. In our 0.6μm technology, the minimum nodal capacitance between these gates is 13.5fF (50% gate capacitance, 50% diffusion capacitance). With a minimum interconnect capacitance of 1.5fF, the minimum nodal capacitance rises to 15fF.

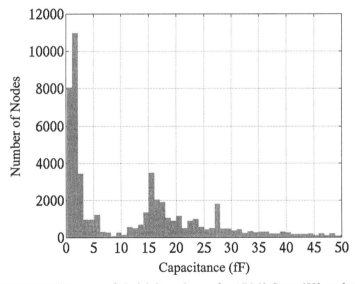

FIGURE 4.2 Histogram of Nodal Capacitance for ARM8 Core. (53k nodes)

Circuit Design Methodology

In Figure 4.2, a histogram of the nodal capacitance of the prototype ARM8 core is shown out to be 50fF. The first peak occurs due to the small diffusion capacitance between the numerous series transistors. These nodes represent internal gate nodes and are not relevant. The next peak, starting around 15fF, represents inter-gate nodes and validates the previous estimate. Nodes in the 10-15fF range occur either for other internal gate nodes, or when the PMOS size has been reduced below the "1x" width of 2.4µm, as will be discussed later

The minimum load capacitance driven by a "1x" gate is 15fF. The test circuit in Figure 4.3 was used to find the maximum load capacitance that a "1x" gate could drive while meeting the MIORFT. The worst case occurs when a driver with a maximum load capacitance drives a gate with minimum load capacitance.

SPICE simulation yielded a C_{MAX} of 50fF. This results in a rise/fall ratio for V_s/V_f of

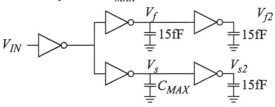

FIGURE 4.3 Test Circuit for Finding C_{MAX} for a 1x Gate Output Driver.

2.85. But what is critical is the input-to-output rise/fall ratio, and since the longer rise/fall time on V_s degrades the rise/fall time on V_{s2} the rise/fall ratio of V_s/V_{s2} is 1.9, which is below the MIORFT of two.

For larger load capacitances, the driver transistor sizes are just scaled up proportionally. A "2x" gate can drive 50-100fF while meeting the MIORFT constraint; a "3x" driver can drive 100-150fF. While a "3x" driver could drive 45-150fF and still meet the MIORFT, the finer resolution on the bins helps to further minimize energy consumption. By using a "2x" instead of a "3x" to drive a 100fF load, the combined capacitance of transistor parasitics and output load has been reduced 10%.

Critical Paths

The timing verification methodology in Section 5.6 is used to identify paths that exceed the target cycle time. Within these paths, gate sizing can be increased to reduce the path delay. The gate delay for the "1x" driver varies by 40% over the range of rated load capacitance. Gates can be sized up to significantly reduce delay at the expense of increased energy consumption. Once the target cycle time has been met, with some headroom, further size increases are not necessary. Since the number of paths that are critical and need to be resized are small, the overall increase in chip energy consumption is insignificant.

To prevent paths from arising that cannot be resized to meet the target cycle time, a maximum logic depth constraint is imposed on the schematic design. This logic depth can be calculated by finding the maximum number of minimum sized inverters in series in which the delay through them is below the target cycle time by some headroom margin. For a 10ns (at 3.3V) target cycle time, and including 10% headroom margin, the maximum logic depth is 30 gates (single inversion, e.g. NAND, NOR, AOI, etc.) per half-cycle.

Thus, a schematic can be guaranteed by design that its layout implementation can meet the target cycle time, preventing radical circuit redesign. If a netlisted schematic has paths with logic depths greater than 30, then the circuit must be redesigned through logic compaction or architectural modification to reduce the logic depth to the allowed amount.

Non-Critical Paths

Paths that have very little logic depth can be made as slow as reasonably possible without impacting target cycle time. However, to minimize short-circuit current, gates with drive strength larger than "1x" are not candidates for size reduction.

Simple gates within a more complex cell, such as an adder or flip-flop, often have minimal capacitive loading. Hence, the PMOS width can be reduced from the nominal 2.4µm down to 1.2µm without exceeding the MIORFT while decreasing the gate-oxide and diffusion capacitance by roughly 33%.

Standard cells are not good candidates for size reduction because their output loading is not known until after place & route, and can change with subsequent re-routes. Custom datapath cells, however, make excellent candidates because the internal loading is known at the time of cell creation. Thus, down-sizing of transistors is done only within custom datapath cells.

4.1.3 Gated Clocks

Gating, or selectively enabling, clocks is critical for energy-efficient circuit implementations, and this is particularly true for general-purpose microprocessors, in which the clocked elements typically require activation only a fraction of the time.

Circuit Design Methodology

The clock drivers described in Section 5.3.2 contain inputs for both a local and global clock enable signal, which allows either entire sections of the processor (e.g. processor core, cache, etc.) to be halted with a global signal, or fine-grained control with a local signal. .

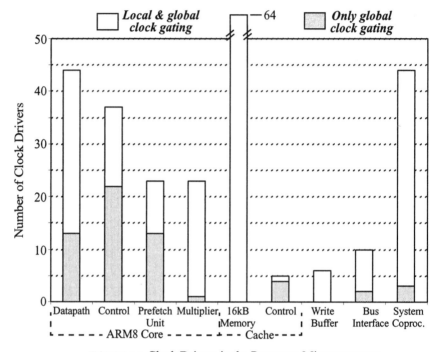

FIGURE 4.4 Clock Drivers in the Prototype Microprocessor

The local enable signal is used whenever the necessary condition for clocking a latch can be calculated from locally available control signals. Routing additional wires across the processor to provide the necessary state information, and adding a large amount of additional logic to calculate the local enable signal can be less energy efficient than always clocking the latch while the global enable signal is asserted. Thus, for latches where the necessary state information is not readily available, the energy penalty for providing this information must be evaluated and compared with the energy savings of having a local clock enable signal. In the prototype system, a total of 256 clock drivers were distributed across the chip as shown in Figure 4.4. Of these, 80% have a local enable signal, demonstrating that fine-grained clock gating can be

General Energy-Efficient Circuit Design

utilized quite extensively in the processor design. Within the processor core itself, 60% of the drivers are locally enabled

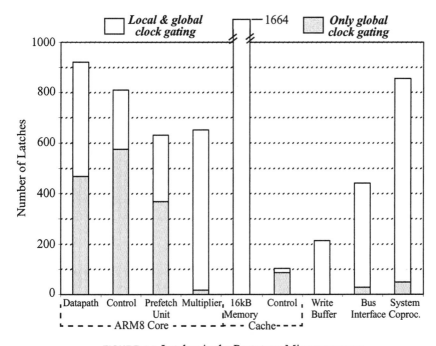

FIGURE 4.5 Latches in the Prototype Microprocessor

These clock drivers, in turn, drive a total of 6292 latches distributed across the processor chip as shown in Figure 4.5. This number does not include memory elements in the register files and in the cache memory. If these latches were clocked every cycle the processor is active, the aggregate clock load would be 150 pF/cycle, which is one-half of the entire processors capacitance/cycle while it is active. However, 75% of these latches were driven with a locally-enabled clock driver, reducing the average, aggregate clock load to somewhere in the 50-75 pF/cycle range. Unfortunately, an exact number is difficult to quantify.

Finally, some latches must be clocked every cycle, even while the processor is completely halted, and they are required in any block that interfaces with the external world (e.g. interrupt controller, memory controller, etc.). Since these latches contribute to the idle power dissipation, it is important to keep the number to a bare mini-

Circuit Design Methodology

mum. In the prototype design, 60 latches required latching every cycle, and contributed 10μW to the idle power dissipation, which is on the same order of magnitude as the subthreshold leakage current power dissipation.

4.1.4 Optimizing Interconnect

The metal profile for our 0.6μm process is shown in Figure 4.6, along with the capacitive components of a representative, minimum-width *Metal2* wire. In this process technology, minimum-width wires have almost a square profile, and as process technology continues to advance, the height of the wires will become significantly greater than their width.

The total capacitance on *Node* is:

$$C_{TOTAL} = C_{top} + C_{bot} + 2 \cdot C_{line} \qquad \text{(EQ 4.2)}$$

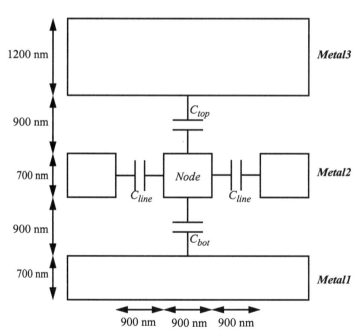

FIGURE 4.6 Interconnect Dimensions and Capacitance Components

88   *Energy Efficient Microprocessor Design*

where the line capacitance, C_{line}, accounts for only 11% of C_{TOTAL} with *Metal1* and *Metal3* present, but 43% in the absence of *Metal1* and *Metal3* as shown in Figure 4.7. In areas of the chip with dense signal routing on all layers, spacing *Metal2* wires at twice-minimum spacing can reduce line capacitance by 11%, or more, depending upon how much *Metal1* and *Metal3* is present around the wire. In regions of the chip loosely populated with wire routes, spacing wires far apart can provide a significant reduction of almost 2x in energy consumption. This is particularly true for *Metal2* and *Metal3* wires, as they are farther from the substrate than *Metal1*.

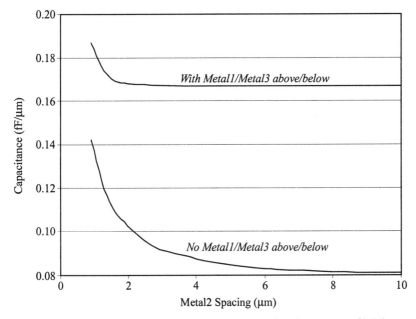

FIGURE 4.7 *Metal2* Wire Capacitance/μm With Adjacent *Metal2* Wires.

In the prototype microprocessor chip, this technique was used pervasively to reduce energy consumption. *Metal3* feedthru wires over the datapath were spaced equidistantly to minimize their overall capacitance. The channel routes in the ARM8 core and in the cache system, in sparsely populated regions, were also spaced farther apart. To minimize the load on the global clock, which transitions every single cycle and is by far the highest energy-consuming net, it was routed in *Metal3* with at least 10μm to the nearest *Metal3* wire. *Metal2* and *Metal1* wires were only utilized to cross underneath and perpendicular to the clock net.

Circuit Design Methodology

In more advanced process technologies, the fraction of C_{LINE} to C_{TOTAL} goes up, which just exacerbates the benefit of spacing wires farther apart than minimum spacing. Copper wires reduce the height of the metal wires, but this height reduction is much less than the lateral geometry shrink going from our 0.6μm process to a much more advance 0.18μm copper process technology.

4.1.5 Layout Considerations

Many layout optimizations that are done for performance improvement or silicon-area efficiency also improve circuit energy efficiency. For example, the layout constraints of the custom datapath cells and the standard cells were carefully optimized to minimize the silicon area in our 3-metal 0.6μm process technology, and by doing so, the overall circuit energy efficiency was increased.

Fingering devices can be used to reduce drain capacitance to not only speed up circuit performance, but reduce the energy consumption. To further reduce drain capacitance, pass gate diffusion can be merged with a driver's diffusion region. Spacing control signals that frequently transition, such as clock signals, away from other cell geometries reduces energy consumption as well.

Datapath Cell Layout

The datapath cell pitch was not set until the entire schematic of the ARM8 core datapath was complete, so that the absolute minimum number of cell feedthrus could be calculated. The initial design yielded thirteen feedthrus, which after schematic redesign, was reduced to ten feedthrus. In addition, further constraints had to be specified to optimize the layout with only three metal layers available for routing.

Within the cell, as shown in Figure 4.8, vertical *Metal1* wires were used to route power and ground. They have higher resistance than *Metal3* wires, but require many less contacts to connect to the devices, and also maximizes capacitance on the power/ground lines. Their width was dictated by the maximum current the cell could draw (Section 5.4.2). *Metal2* was utilized for control lines that span the entire datapath, as well as local cross-overs of the *Metal1* power/ground lines. They were spaced as far apart as possible to reduce parasitic capacitance on them. *Metal3* was used exclusively for feedthrus. The minimum pitch of *Metal3* to accommodate contacts is 2.55μm, so in order to allocate room for ten feedthrus, the cell height was set to 25.5μm. These constraints minimized the overall area of the datapath, which in turn minimized the length of the long feedthrus across the datapath, and reduced their capacitance and energy consumption, as well.

General Energy-Efficient Circuit Design

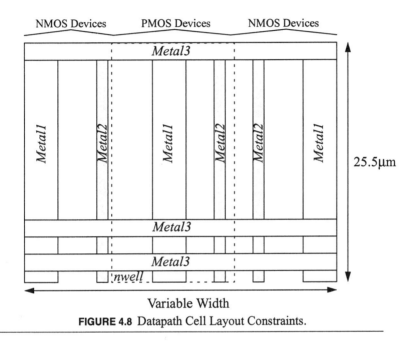

FIGURE 4.8 Datapath Cell Layout Constraints.

The cells were designed so that they can abut on top and bottom. which allowed them to be tiled up to form the datapath. By placing the power/ground lines on the far left and right of the cell, it can also directly abut other datapath cells on either side if a routing channel is not required. In better process technologies with more metal layers available, the cells can always abut because the additional metal layers remove the need for explicit routing channels.

Standard Cell Layout

In designing the standard cell library, the goal was to minimize the overall area of synthesized layout, which was achieved by reducing the cell size as small as possible. The pitch was set to 19.2μm, as shown in Figure 4.9, which allowed for a twice-minimum size PMOS device to be placed without having to finger it. To free up as much *Metal2* as possible for the router, no *Metal2* was allowed inside the cell for routing, and all pins had to be placed, centered about the middle in *Metal2*, on a 2.4μm routing pitch. This allowed the router to use *Metal2* over the cell. The router used *Metal3* hor-

Circuit Design Methodology

izontally over the cell, *Metal1* horizontally outside the cell, and *Metal2* for vertical routes. The cells had to be designed to abut on either side.

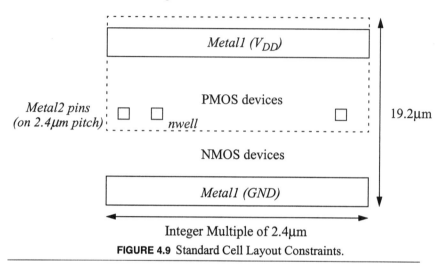

FIGURE 4.9 Standard Cell Layout Constraints.

4.2 Memory Design

The basic memory blocks used in both the cache memory and the external SRAM have been derived from a previous design which utilized sub-blocking, self-timing, and charge-sharing sense-amplifiers for a very low-energy implementation *[4.5]*. In addition, this design was extended to a CAM which was utilized for the cache tags. The key modification made to the previous design to make it DVS compatible was changing the charge-sharing sense-amplifier, which uses an NMOS pass gate to limit the signal swing on the bitlines to $V_{DD} - V_T$, to a full-swing design.

The critical aspect of a memory design, in order to ensure DVS compatibility, are those circuits which are not standard CMOS logic. These primarily include the memory cell, which only pulls down the bitline voltage by some fraction of V_{DD}, and the sense-amp circuit which restores the signal on the bitlines to full-scale. While allowing the bitlines to swing full-scale would improve circuit robustness for DVS, this would significantly increase memory energy consumption and delay, and is therefore not a viable option.

Memory Design

The rest of the memory circuits (i.e. address decoder, word-line driver, output buffer, and control circuitry) are typically implemented with standard static or dynamic CMOS logic, for which the circuit delay scales with voltage similarly to any other logic circuits.

4.2.1 SRAM

The critical part of the SRAM's signal path along the bitlines, including the memory cell and sense-amplifier circuits, is shown in Figure 4.10. The width of the sense amplifier layout is twice that of the memory cell, so a 2-to-1 multiplexer is used for column decoding to provide efficient, compact layout by pitch-matching the sense amplifier to two memory cells. CMOS pass gates are required to implement the bi-directional multiplexer.

The bitlines on either side of the multiplexer are precharged to V_{DD}, so while the SRAM is not being actively accessed, these precharged nodes will vary in voltage with V_{DD}. The internal state of each memory cell, which is maintained by the cross-coupled inverters, will also scale in voltage as V_{DD} varies. Thus, when the SRAM is inactive, it can tolerate transient variations on V_{DD} much like static CMOS circuits because all logic high nodes are actively being pulled up by a PMOS device.

Writing to the SRAM cell requires pulling one of the bitlines all the way to ground in order to flip the state of the cell's cross-coupled inverters. This is accomplished by one of the NMOS pull-down devices on *Bit* and $\overline{Bit}$, which is enabled by its corresponding NOR gate when the write enable (*Wen*) signal is high. The delay of this signal path will scale with varying V_{DD} much like static CMOS logic.

When the SRAM cell is being read from, the cross-coupled inverter, whose output is low, begins to pull down one of the bitlines through the NMOS pass-gate activated by the *Word* signal. Both NMOS devices are minimum size to reduce the size of the SRAM cell (which determines the total SRAM block size) and therefore can only pull the bitline down slowly. The sense amplifier is used so that only some fraction of the voltage V_{DD} has to be developed across the bitlines to register a signal transition.

The memory is self-timed, which in addition to minimizing switching activity to significantly reduce energy consumption, also enables the delay of the SRAM read to scale with varying V_{DD} similar to static CMOS logic. A dummy word line is used to generate the *Sense* signal, which is delayed from the activation of the *Word* signal by:

Circuit Design Methodology

Circuit Design Methodology

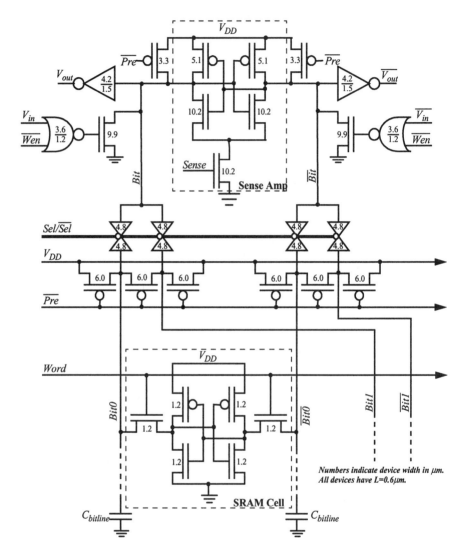

FIGURE 4.10 SRAM Cell and Sense Amp.

$$t_{Word \rightarrow Sense} = \frac{2 \cdot C_L \cdot V_{DD}}{I_1(V_{DD})} \quad \text{(EQ 4.3)}$$

which is just the delay through two static CMOS gates. The voltage differential generated on the bitline at the input of the sense-amp when it is activated is:

$$\Delta V_{Bit|0} = \frac{I_2(V_{DD}) \cdot t_{Word \rightarrow Sense}}{C_{Bitline}} = \frac{I_2(V_{DD}) \cdot C_L \cdot V_{DD}}{I_1(V_{DD}) \cdot C_{Bitline}} = \alpha \cdot V_{DD} \quad \text{(EQ 4.4)}$$

where α is a voltage-independent term because the voltage-dependence in the ratio of the current terms, I_1 and I_2, cancels out in Equation 4.4. Thus, the voltage drop is proportional to V_{DD}. The delay from the activation of the *Sense* signal, at which point the voltage on *Bit* is $\Delta V_{Bit|0}$, to the *Bit* signal crossing $V_{DD}/2$ so that a signal transition registers on V_{out}, is:

$$t_{Sense \rightarrow Bit} = \frac{C_{Bit} \cdot \Delta V_{Bit}}{I_{sa}(V_{DD})} = \frac{C_{Bit} \cdot \left((V_{DD} - \Delta V_{Bit|0}) - \frac{V_{DD}}{2}\right)}{I_{sa}(V_{DD})} = \frac{C_{Bit} \cdot V_{DD} \cdot (0.5 - \alpha)}{I_{sa}(V_{DD})}$$

where I_{sa} is the average current in the sense-amp's series NMOS transistors as *Bit* varies from $V_{DD} - \Delta V_{Bit|0}$ to $V_{DD}/2$, and scales with V_{DD} similar to a static CMOS gate.

Thus, with self-timing, the voltage differential generated at the input of the sense-amp is proportional to V_{DD}, which then allows the delay of the signal path from *Word* to V_{out} to scale with V_{DD} similar to static CMOS logic. This is demonstrated in Figure 4.11 which plots this delay versus static CMOS logic over V_{DD}.

So while this delay tracks well for constant V_{DD}, when V_{DD} dynamically varies while the sense-amp is evaluating, this delay begins to deviate. This occurs because the voltage on *Bit* remains independent of V_{DD} during the sensing, so that while the voltage differential on *Bit* to flip the sense-amp at constant voltage is proportional to V_{DD}:

$$\Delta V_{Bit} = (V_{DD} - \Delta V_{Bit|0}) - V_{DD}/2 \quad \text{(EQ 4.5)}$$

when V_{DD} varies by ΔV_{DD}, the required voltage differential scales inversely with ΔV_{DD}:

$$\Delta V_{Bit}(\Delta V_{DD}) = (V_{DD} - \Delta V_{Bit|0}) - (V_{DD} - \Delta V_{DD})/2 = \Delta V_{Bit} + (\Delta V_{DD})/2 \quad \text{(EQ 4.6)}$$

Circuit Design Methodology

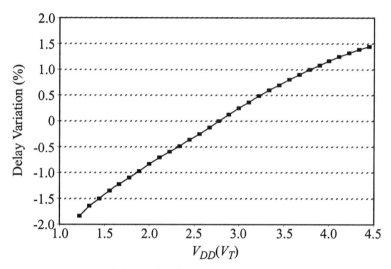

FIGURE 4.11 Relative Delay from *Word* to V_{out} vs. Static logic for Constant V_{DD}.

Thus, when ΔV_{DD} is positive, indicating that V_{DD} is falling, the amount of voltage required to switch, $\Delta V_{Bit}(\Delta V_{DD})$, actually increases with V_{DD} and causes the sense-amp to slow down much faster than static CMOS logic. Likewise, when ΔV_{DD} is negative, indicating that V_{DD} is rising, the value $\Delta V_{Bit}(\Delta V_{DD})$ actually decreases with V_{DD}, and causes the sense-amp to speed up much faster than static CMOS logic. As shown in Section 4.5.3, this issue is most critical at low V_{DD}, and ultimately limits how fast V_{DD} can be allowed to vary. This is a fundamental limitation of sense-amps.

4.2.2 CAM

A traditional implementation of a CAM cell is shown in Figure 4.12, which uses a dynamic NOR gate (*M3*) to generate a *Match* signal that remains high if the data values placed on the *Bit*/$\overline{Bit}$ lines completely matches the cells' contents across the entire row [4.2]. If any one bit in the row mismatches against the input pattern (i.e. *Bit* ≠ *m* and $\overline{Bit}$ ≠ $\overline{m}$), one of the NMOS pass gates, *M1* or *M2*, pulls the input to *M3* high, which in turn pulls *Match* low.

This implementation is not DVS compatible, due to the NMOS pass gates. At least two more PMOS transistors are required to convert them to CMOS pass gates. Also,

Memory Design

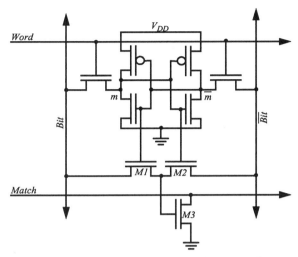

FIGURE 4.12 Traditional CAM Cell.

the traditional CAM cell requires Bit and $\overline{Bit}$ to be pre-discharged for a match operation, and is not compatible with read and write operations which require Bit and $\overline{Bit}$ to be pre-charged.

The CAM cell was redesigned, as shown in Figure 4.13, so that the CAM always begins an operation with Bit and $\overline{Bit}$ pre-charged. This eliminates the delay and unnecessary energy cost of switching the polarity on the bitlines. In addition, the CAM cell is now DVS compatible.

The additional five transistors increased the size of the memory cell so that the sense-amp was better pitch-matched to a single memory cell in the CAM, as compared to two memory cells in the SRAM. This change eliminated the need for a column multiplexer so that the bitlines directly drive the sense amp. However, this forces the bitlines to swing full-rail upon a CAM read. Since the CAM is most commonly performing match operations, thereby making CAM reads infrequent, the overall increase in system energy consumption is negligible.

The CAM memory utilizes the same sense-amp and self-timed circuitry as the SRAM, so that its delay has the same characteristics as the SRAM. The delay tracks static CMOS logic very well at constant voltage, but suffers the same deviation when V_{DD} varies during the sensing period.

Circuit Design Methodology

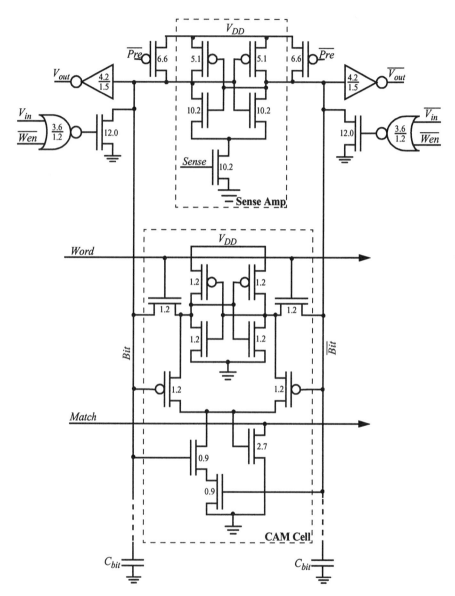

FIGURE 4.13 CAM Cell and Sense Amp.

4.2.3 Register File

The ARM8 architecture requires a 30x32b register file with one write-port and two read-ports. The three ports are independently operated, requiring three different ports to the register cell itself. To reduce the requisite routing overhead, the ports were implemented with single-ended pull-down circuits, as shown in Figure 4.14. The single-ended ports forced both the input and output bitlines to swing full-rail, but this ensures that the delay of the register file scales over V_{DD} similar to static CMOS logic.

Both *BitA* and *BitB* are pre-charged high, and are selectively pulled down if both the cell's read signal (*ReadA* or *ReadB*) is high and the internal cell state is high. The

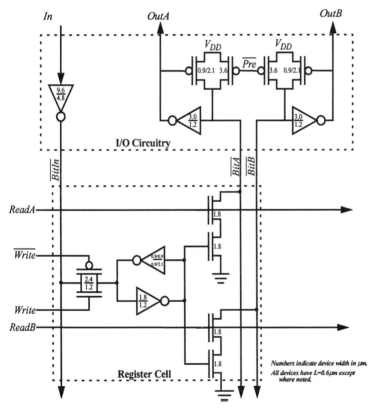

FIGURE 4.14 Register File Cell and I/O Circuits.

Circuit Design Methodology

Circuit Design Methodology

input data is inverted, because simulation demonstrated that the majority of data bits written to the register file are low. This reduced the energy consumption of the register file datapath by 55%, and the overall register file energy consumption by 33%. The weak feedback transistors are required to maintain state on *BitA* and *BitB* when it is not actively being pulled down in the evaluation state.

4.3 Low-Swing Bus Transceivers

The energy required to drive large busses has become increasingly significant as process technology has improved. While average gate capacitance and local interconnect capacitance have reduced with improved process technology, global bus capacitance has not. Global busses include both intrachip busses, and the interchip, board-level processor bus.

Global intrachip bus capacitance has actually increased with improved process technology, because both wiring capacitance per unit length and average bus length increase with improved process technology. The wiring capacitance grows due to thinner oxides and narrower wiring pitches. Larger die sizes necessitate larger bus lengths for connecting the various microprocessor blocks. The deployment of copper interconnect has enabled a reduction in the wiring capacitance per unit length, because it could be manufactured thinner than more traditional aluminum interconnect while maintaining the same resistivity. But as copper interconnect is pushed into more advanced process technologies, the wiring capacitance per unit length will continue to once again increase.

Interchip busses, primarily the external process bus, have a capacitance that has remained roughly constant, because it is dominated by printed circuit board capacitance, and packaging parasitic capacitance. However, bus frequencies have greatly increased, driving up the energy consumption due to fast signal edges.

Transceivers for both intrachip and interchip busses are presented, along with measured results from a test chip. These transceivers were not integrated into the prototype system, in order to aid debugging, but integration into a future processor system is discussed. The transceivers were designed to be DVS compatible so that they could be integrated into a DVS processor system.

The key enabler of these low-swing transceivers is the demonstration of a high-efficiency, low-voltage regulator *[4.6]*. The output voltage can be as low as 200mV with a conversion efficiency in excess of 90% using a standard 3.8V lithium-ion battery for the input voltage.

4.3.1 Intrachip Transceivers

The on-chip transceiver was designed with differential signal lines. This eliminates the need for a reference voltage at the receiver, and makes the bus significantly more immune to coupling from adjacent, interfering signal lines. The penalty is that the bus requires twice as many signal routes, but as the number of metal layers continues to increase with process technology, this penalty is mitigated.

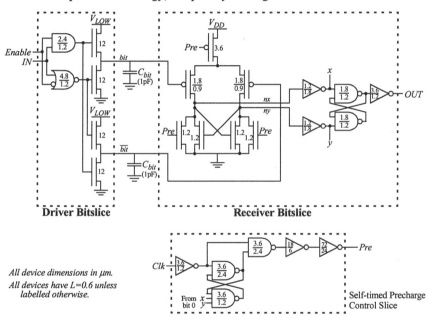

FIGURE 4.15 Differential Low-Swing Bus Transceiver Circuit

The transceiver architecture is shown in Figure 4.15. The driver uses two pairs of two NMOS devices to drive the bus signals *bit* and $\overline{bit}$ to either V_{LOW} or ground. Since the driver's NAND and NOR enabled gates are powered by the variable voltage V_{DD}, the driver current and delay scales with V_{DD}. The voltage on *bit* and $\overline{bit}$ never exceeds V_{LOW}, which can be little as 200mV, so that the receiver requires a comparator with PMOS inputs to sense the voltage difference. To minimize energy consumption, the

Circuit Design Methodology

comparator utilizes a clocked, dynamic design. The self-timed precharge signal simplifies the receiver control, which only requires a single clock signal to control it.

The $\overline{RS}$ latch at the output of the comparator registers when one of the comparator's outputs, nx or ny, has gone high, and changes the signal OUT accordingly. This eliminates any spurious transitions on the output signal. Since the comparator and the $\overline{RS}$ latch is powered by V_{DD}, the delay through the receiver scales with V_{DD} as well, so that the transceiver delay and energy consumption scales with V_{DD} for DVS compatibility. The self-timed precharge circuit puts the comparator back into precharge mode as soon as a transition is detected in order to minimize energy consumption.

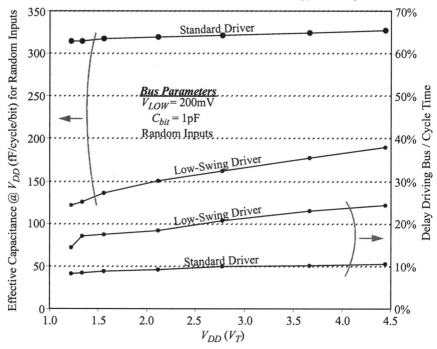

FIGURE 4.16 Low-Swing vs. Standard Driver for Intrachip Busses (Simulated).

Figure 4.16 compares both the delay and energy consumption (in capacitance/cycle) of the low-swing bus transceiver and a standard static CMOS bus transceiver when driving a 1pF load. The capacitance/cycle was calculated for random inputs, which yields a 50% probability of a signal transition. While the low-swing transceiver delay is twice as long, the capacitance/cycle reduction varies from 1.7x at high voltage to over 2.5x at low voltage.

Thus, if architectural design can hide this increased bus latency and place a latch at the receiver, the low-swing intrachip transceiver can provide significant reduction in energy consumption. Additionally, for larger bus capacitances, the capacitance reduction will increase. A 2pF bus capacitance will double the capacitance/cycle of the standard bus transceiver, but the low-swing transceiver will only increase by 30% at low voltage, and less than 2% at high V_{DD}. This occurs because the bulk of the energy consumed in the low-swing transceiver is by the comparator.

One other limitation to the low-swing transceiver is that while a standard bus driver will only consume energy upon an input signal transition, the low-swing transceiver consumes energy independent of the input signal. Thus, if the data is highly correlated, the low-swing transceiver will actually consume more energy than the standard bus driver. For the 1pF bus capacitance, if the probability of a transition drops from 50% to 20%, then the standard bus driver becomes more energy-efficient. So in order to evaluate whether a low-swing transceiver can reduce the energy consumption on a bus, both its capacitance and data correlation must be evaluated.

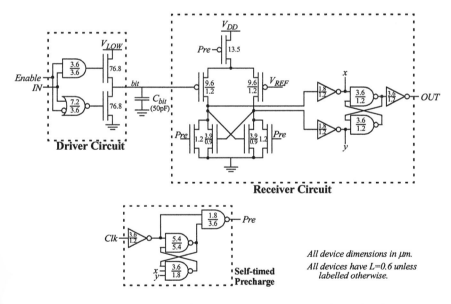

FIGURE 4.17 Single-ended Low-Swing Bus Transceiver Circuit.

4.3.2 Interchip Transceivers

The off-chip transceiver is a non-differential version of the on-chip transceiver. While the differential signaling provides more robustness, the additional pin-count on the package cannot be tolerated. Thus, the second bitline was removed, and a reference voltage, V_{REF}, set to one-half of V_{LOW} is used as the second input on the differential comparator. The modified transceiver architecture is shown in Figure 4.17. The single-ended receiver does place a constraint on the minimum V_{DD} value:

$$V_{DD} > V_{REF} + V_{Tp} = \frac{V_{LOW}}{2} + V_{Tp} \qquad \text{(EQ 4.7)}$$

so that as V_{LOW} is increased to provide more circuit robustness, the trade-off is decreased operating range on V_{DD}.

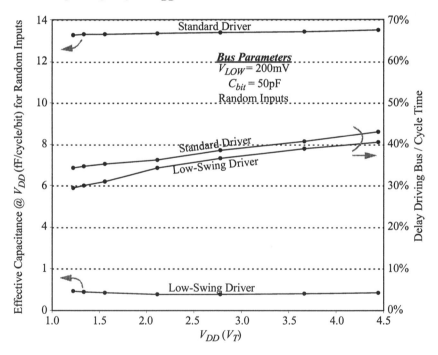

FIGURE 4.18 Low-Swing vs. Standard Driver for External Busses (Simulated).

Low-Swing Bus Transceivers

Figure 4.18 compares both the delay and energy consumption (in capacitance/cycle) of the single-ended low-swing bus transceiver and a standard static CMOS bus transceiver when driving a 50pF load. This is a typical capacitance for a bus with ten external chips connected to it, as in the prototype processor system. The capacitance/cycle was calculated for random inputs, which yields a 50% probability of a signal transition. For this case, the low-swing transceiver not only provides slightly less path delay across the bus, but a significant reduction in capacitance/cycle in excess of 15x. Correlation in the bus data will reduce the margin of savings, but the probability of a signal transition would have to be below 3.3% before the low-swing transceiver becomes less energy-efficient. Thus, the low-swing transceiver is extremely well-suited for the external PCB processor bus, which has the key characteristic that the capacitance per bit is anywhere from 25-100pF.

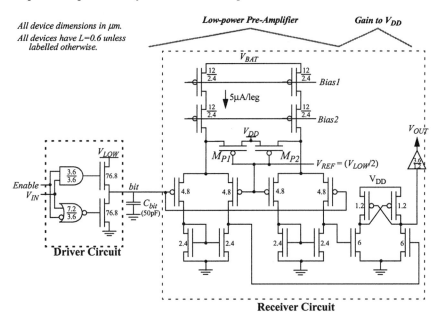

FIGURE 4.19 Single-ended Low-Swing Clock Transceiver Circuit.

A clock signal is still required to drive the external bus, and each cycle it is required to switch on the order of 50pF of capacitance. Switching the signal at V_{DD} will cause the energy consumed by the clock signal to completely dominate the energy con-

Circuit Design Methodology 105

Circuit Design Methodology

sumed by the low-swing bus. Thus, a continuous transceiver was developed to transmit the clock signal at low voltage to mitigate this component of energy consumption, and this circuit is shown in Figure 4.19.

The driver is the same circuit used in the previous transceivers, and converts the input clock signal to a low-swing signal. The receiver consists of two components, the pre-amplifier and the second gain stage. The pre-amplifier's dual source-coupled pair (SCP) circuits convert the signal to differential, and amplifies the signal from $0-V_{LOW}$ to $0-V_{tp}$. The SCP circuits are biased by the battery voltage, V_{BAT}, to ensure a fixed minimum tail-current of 10μA per SCP, which is set by bias voltages, $Bias1$ and $Bias2$, from a high-swing cascode current source [4.7]. The devices M_{P1} and M_{P2} provide an additional current source, which is a function of the variable voltage, V_{DD}. This allows the speed, and the energy consumption of the pre-amplifier to scale with V_{DD}. The second-gain stage amplifies the clock signal to $0-V_{DD}$. The cross-coupled loads ensure that this gain stage has no static current, as only one of this gain stage's NMOS devices will be turned on, since $V_{tp} > V_{tn}$. Thus, by using the pre-amplifier to minimize short-circuit current, the receiver provides the necessary signal level-conversion with minimal energy consumption.

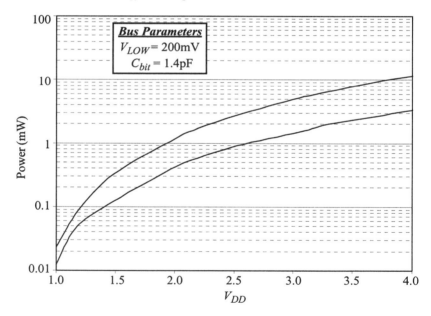

FIGURE 4.20 Power of Low-Swing and Standard Driver for Intrachip Busses.

The energy consumption in terms of the effective switched capacitance/cycle at V_{DD} is shown in Figure 4.21. A clock signal switching at a voltage V_{DD} would have 50 pF/cycle, while the low-swing clock transceiver has reduced this to less than 2 pF/cycle. The capacitance/cycle is roughly constant, due to the variable-current tail-source in the pre-amplifier. The penalty of using the continuous-time clock transceiver, rather than the previous transceiver with the dynamic latch, is approximately 2x.

4.3.3 Test Chip

A test chip was fabricated in our 0.6μm process to validate the low-swing transceiver designs. The intrachip receiver successfully operates with V_{LOW} = 200mV as V_{DD} ranges from 1.0-4.2V, and the clock frequency ranges from 4-111MHz. For the 1.4pF bus (measured) on-chip bus, the reduction in power dissipation ranges from 1.7x to 3.3x as demonstrated in Figure 4.20, which yields an equivalent reduction in energy consumption. In addition, the intrachip transceiver can operate as V_{DD} varies at a rate of 10 V/μs, demonstrating that this design is compatible with a DVS processor sys-

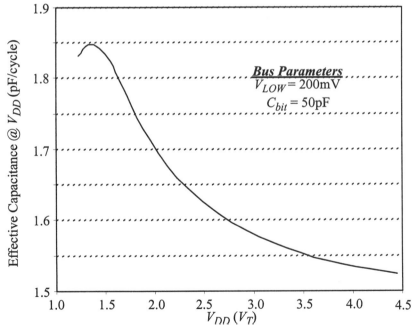

FIGURE 4.21 Low-Swing Clock Transceiver Energy Consumption (Simulated).

tem. At $V_{DD} = 1.0$V, V_{LOW} can be operated as low as 40mV, though the minimum V_{LOW} for all values of V_{DD} is 150mV.

The inter-chip receiver successfully operates on a 50 pF/bit bus with V_{LOW} of 200mV as V_{DD} ranges from 1.0-3.5V, and the clock frequency ranges from 4-100MHz for the worst-case condition when all bits are switching simultaneously. Higher values of V_{DD} increase the current draw from V_{LOW}, and the resulting noise prevents the receiver from continuing to operate properly. If V_{LOW} is increased to 500mV, the receiver can operate over the range 1.25-3.75V. When the number of bits that switch simultaneously is reduced the receiver can operate at a value of V_{DD} as high as 4.75V due to the decreased current draw on V_{LOW}.

In addition, when all bits are simultaneously switching, V_{LOW} is 500mV, and V_{DD} varies at a rate of 16 V/μs, the receiver can operate over the V_{DD} range 1.25-3.25V. As V_{LOW} decreases to 200mV, the range of V_{DD} drops to 1.0-2.6V. Decreasing dV_{DD}/dt to 1 V/μs allows proper operation over the same range of V_{DD} as when it is held constant. These results are summarized in Figure 4.22.

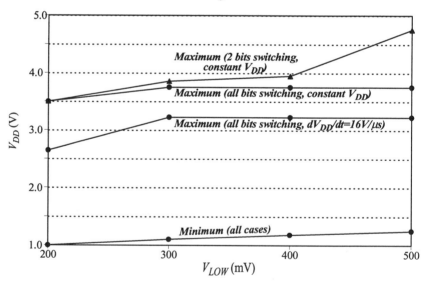

FIGURE 4.22 Range of Operation for Low-Swing Interchip Bus.

The prototype low-swing inter-chip bus was designed with one power/ground pin per eight bus pins. It is believed that by decreasing this ratio the range of operation for

V_{DD} can be increased. In addition, further on-chip bypass capacitance for V_{LOW} will reduce the on-chip noise of this signal, and also improve circuit robustness and operating range.

4.3.4 Further Integration

The prototype chip demonstrates the viability of the low-swing bus transceivers, as well as significant reductions in energy consumption, particularly for the external PCB processor bus. Extra overhead is required for the additional voltage regulator to generate V_{LOW}, but this can mitigated by sharing circuits between this regulator and the existing system voltage regulator [4.6].

It is believed that additional pins for the low-voltage power and ground will alleviate the headroom problems observed on the test chip, but this requires further investigation. Additionally, for the inter-chip transceiver, it is necessary to provide a stable V_{REF} signal on-chip. For the test-chip, V_{REF} was generated externally with a resistor divider. However, this needlessly dissipates static power dissipation. Further investigation is required to minimize this static power. One possible solution is to generate V_{REF} via a switched capacitor network. Another solution is to use large bypass capacitors either on-chip or internal to the package to maximize the size of the resistors used in the divider.

4.4 Design Constraints Over Voltage

A typical processor targets a fixed supply voltage, and is designed for ±10% maximum voltage variation. Correct functional operation must be verified over this small voltage variation, as well as a slew of timing constraints, such as maximum path delay, minimum path delay, maximum clock skew, and minimum noise margin. In contrast, a DVS processor (Section 2.5) must be designed to operate over a much wider range of supply voltages, which impacts both design implementation and verification time.

4.4.1 Circuit Design Constraints

To realize the full range of DVS energy efficiency, only circuits that can operate all the way down to V_T should be used. NMOS pass gates are often used in low-power design due to their small area and input capacitance. However, they are limited by not being able to pass a voltage greater than $V_{DD} - V_{Tn}$, such that a minimum V_{DD} of $2 \cdot V_T$ is required for proper operation. Since throughput and energy consumption vary by 4x over the voltage range V_T to $2 \cdot V_T$, using NMOS pass gates restricts the range of operation by a significant amount, and are not worth the moderate improvement in energy efficiency. Instead, CMOS pass gates, or an alternate logic style, should be utilized to realize the full voltage range of DVS.

As previously demonstrated in Figure 2.14, the delay of CMOS circuits track over voltage such that functional verification is only required at one operating voltage. The one possible exception is any self-timed circuit, which is a common technique to reduce energy consumption in memory arrays. If the self-timed path layout exactly mimics that of the circuit delay path as was done in the prototype design, then the paths will scale similarly with voltage and eliminate the need to functionally verify over the entire range of operating voltages.

4.4.2 Circuit Delay Variation

While circuit delay tracks well over voltage, subtle delay variations exist and do impact circuit timing. To demonstrate this, three chains of inverters were simulated whose loads were dominated by gate, interconnect, and diffusion capacitance respectively. To mimic paths dominated by stacked devices, a fourth inverter chain was simulated in which both the PMOS and NMOS transistors were each source-degenerated with an additional three equally-sized series transistors, effectively modeling a four-transistor stack. The relative delay variation of these circuits is shown in Figure 4.23 for which the baseline reference is an inverter chain with a balanced load capacitance similar to the ring oscillator.

The relative delay of all four circuits is a maximum at only the lowest or highest operating voltages. This is true even including the effect of the interconnect's RC delay. Since the gate dominant curve is convex, combining it with one or more of the other effects' curves may lead to a relative delay maxima somewhere between the two voltage extremes. However, all the other curves are concave and roughly mirror the gate dominant curve such that this maxima will be less than a few percent higher than at either the lowest or highest voltage, and therefore insignificant. Thus, timing analysis is only required at the two voltage extremes, and not at all the intermediate voltage values.

As demonstrated by the series dominant curve, the relative delay of four stacked devices rapidly increases at low voltage. Additional devices in series will lead to an

Design Constraints Over Voltage

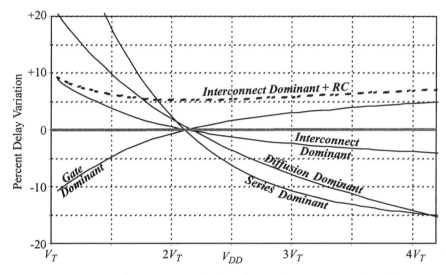

FIGURE 4.23 Relative CMOS Circuit Delay Variation over Supply Voltage.

even greater increase in relative delay. As supply voltage increases, the drain-to-source voltage increases for the stacked devices during an output transition. For the devices whose sources are not connected to V_{DD} or ground, their body-effect increases with supply voltage, such that it would be expected that the relative delay would be a maximum at high voltage. However, the sensitivity of device current and circuit delay to gate-to-source voltage exponentially increases as supply voltage goes down. So even though the magnitude change in gate-to-source voltage during an output transition scales with supply voltage, the exponential increase in sensitivity dominates such that stacked devices have maximum relative delay at the lowest voltage.

Thus, to improve the tracking of circuit delay over voltage, a general design guideline is to limit the number of stacked devices, which was four in the case of the prototype design. One exception to the rule is for circuits in non-critical paths, which can tolerate a broader variation in relative delay. By using the clocking methodology described in Section 5.3, it can be ensured that this broader variation does not lead to potential race conditions. Another exception is for circuits whose alternative design would be significantly more expensive in area and/or power (e.g. memory address decoder), but the circuits must still be designed to meet timing constraints at low voltage.

Circuit Design Methodology

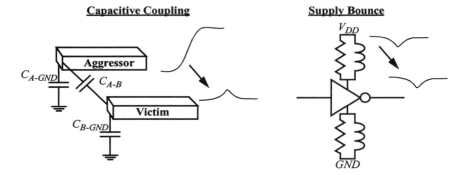

FIGURE 4.24 Noise Margin Degradation.

4.4.3 Noise Margin Variation

Figure 4.24 demonstrates the two primary ways that noise margin is degraded. The first is capacitive coupling between an aggressor signal wire that is switching and an adjacent victim wire. When the aggressor and victim signals have the same logic level, and the aggressor transitions between logic states, the victim signal can also incur a voltage change. If this change is greater than the noise margin, the victim signal will glitch and potentially lead to functional failure. Supply bounce is induced by switching current spikes on the power distribution network, which has resistive and inductive losses. If the gate's output signal is the same voltage as the supply that is bouncing, the voltage spike transfers directly to the output signal. Again, if this voltage spike is greater than the noise margin, glitching, and potentially functional failure, will occur.

For the case of capacitive coupling, the amplitude of the voltage spike on the victim signal is proportional to V_{DD} to first order. As such, the important parameter to analyze is noise margin divided by V_{DD} to normalize out the dependence on V_{DD}. Figure 4.25 plots two common measures of noise margin vs. V_{DD}, the noise margin of a standard CMOS inverter, and a more pessimistic measure of noise margin, V_T. The relative noise margin is a minimum at high voltage, such that signal integrity analysis to ensure there is no glitching only needs to consider a single value of V_{DD}. If a circuit passes signal integrity analysis at maximum V_{DD}, it is guaranteed to pass at all other values of V_{DD}.

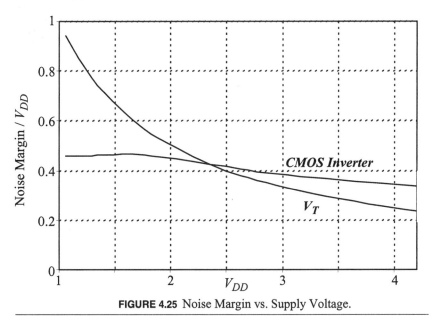

FIGURE 4.25 Noise Margin vs. Supply Voltage.

Supply bounce occurs through resistive (IR) and inductive (dI/dt) voltage drop on the power distribution network both on chip and through the package pins. Figure 4.26 plots the relative normalized IR and dI/dt voltage drops as a function of V_{DD}. It is interesting to note that the worst case condition occurs at high voltage, and not at low voltage, since the decrease in current and dI/dt more than offsets the reduced voltage swing. Given a maximum tolerable noise margin reduction, only one operating voltage needs to be considered, which is maximum V_{DD}, to determine the maximum allowed resistance and inductance. The global power grid and package must then be designed to meet these constraints on resistance and inductance.

4.4.4 Delay Sensitivity

Supply bounce has another adverse affect on circuit performance in that it can induce timing violations. Supply bounce decreases a transistor's gate drive, which in turn increases the circuit delay. If this increase occurs within a critical path, a timing violation may result leading to functional failure.

Circuit Design Methodology

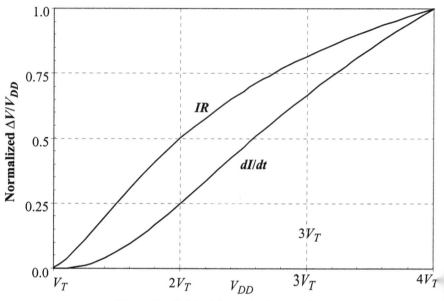

FIGURE 4.26 Normalized Noise Margin Reduction due to Supply Bounce.

A typical microprocessor uses a phase-locked loop to generate a clock frequency which is locked to an external reference frequency and independent of on-chip voltage variation. As such, both global and local voltage variation can lead to timing violations if the voltage drops a sufficient amount to increase the critical paths' delay past the clock cycle time. However, in the DVS system, the clock signal is derived from a ring oscillator whose output frequency is strictly a function of V_{DD}, and not an external reference. As such, global voltage variations not only slow down the critical paths, but the clock frequency as well such that the processor will continue operating properly.

Localized supply variation, however, may only effect the critical paths, and not the ring oscillator. These can lead to timing violations if the local supply drop is sufficiently large. As such, careful attention has to be paid to the local supply routing. For the prototype design, a design margin of 5% was included in the timing verification to allow for localized voltage drops.

Design Constraints Over Voltage

Delay sensitivity is the relative change in delay given a drop in V_{DD}, and can be calculated as:

$$\frac{\partial \text{Delay}}{\text{Delay}}(V_{DD}) = \frac{\partial \text{Delay}}{\partial V_{DD}} \cdot \lim_{\Delta V_{DD} \to 0} \left(\frac{\Delta V_{DD}}{\text{Delay}(V_{DD})} \right) \quad \text{(EQ 4.8)}$$

This equation can be analytically quantified using Equation 4.8, and the normalized delay sensitivity is plotted as a function of V_{DD} in Figure 4.27. For sub-micron CMOS processes, the delay sensitivity peaks at approximately $2 \cdot V_T$. Thus, the design of the local power grid only needs to consider one value of V_{DD}, $2 \cdot V_T$, to ensure that the resistance/inductance voltage drop meets the design margin on delay variation. If the power grid meets timing constraints at this value of V_{DD}, it is guaranteed to do so at all other voltages.

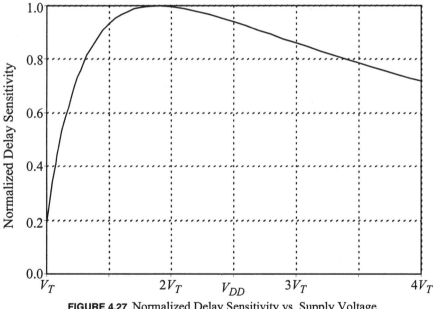

FIGURE 4.27 Normalized Delay Sensitivity vs. Supply Voltage.

Circuit Design Methodology

4.4.5 Summary

The verification complexity and design margins of a DVS-compatible processor are very similar to any other high-performance processor. One added constraint is that the circuits must be able to properly operate from V_T to maximum V_{DD}. Additionally, timing verification is required at both maximum and minimum voltage, instead of just a single voltage.

Since the clock frequency is generated from the on-chip ring oscillator, the constraints on the global power distribution are actually not as severe, because only local voltage drops can induce timing failure, not global voltage drops. The only constraint put upon the global power distribution is to ensure that noise margins are met, which is much less restrictive than designing it to be immune to timing failure with an externally referenced clock signal.

4.5 Design Constraints for Varying Voltage

One approach for designing a processor system that switches voltage dynamically is to halt processor operation during the switching transient. The drawback to this approach is that interrupt latency is increased and potentially useful processor cycles are discarded. However, static CMOS gates are quite tolerable to supply voltage slew, so there is no fundamental need to halt operation during the transient.

For the simple inverter in Figure 4.28, when V_{in} is high the output remains low irrespective of V_{DD}. However, when V_{in} is low, the output will track V_{DD} via the PMOS device, and can be modeled as a simple RC network. In our 0.6μm process, the RC time constant is a maximum of 5ns, at low voltage where it is largest. Thus, the inverter tracks quite well for a dV_{DD}/dt in excess of 200 V/μs.

Because all the logic high nodes will track V_{DD} very closely, the circuit delay will instantaneously adapt to the varying supply voltage. Since the processor clock is derived from a ring oscillator also powered by V_{DD}, its output frequency will dynamically adapt as well, as demonstrated in Figure 4.29.

Thus, static CMOS is well-suited to continue operating during voltage transients. However, there are design constraints when using a design style other than static CMOS.

Design Constraints for Varying Voltage

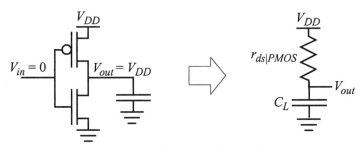

FIGURE 4.28 Equivalent RC Network for Static CMOS Inverter.

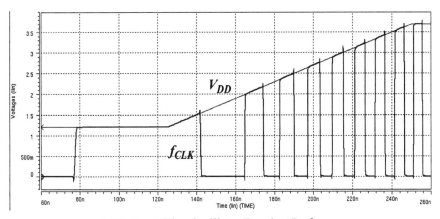

FIGURE 4.29 Ring Oscillator Transient Performance.

4.5.1 Dynamic Circuits

Dynamic logic styles are often preferable over static CMOS as they are more efficient for implementing complex logic functions. They can be used with a varying supply voltage, as long as their failure modes are avoided by design. These two failure modes for a simple dynamic circuit are shown in Figure 4.30, and occur while the circuit is in the evaluation state ($\phi=1$) and V_{in} is low. In this state, V_{out} has been precharged high, and is floating during the evaluation state.

Circuit Design Methodology   **117**

Circuit Design Methodology

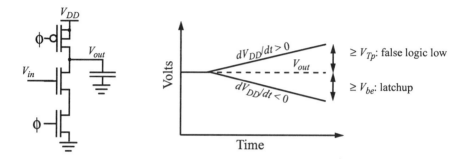

FIGURE 4.30 Failure Modes for Dynamic Logic with Varying V_{DD}.

If V_{DD} ramps down by more than a diode drop, V_{be}, by the end of the evaluation state, the drain-well diode will become forward biased. This current may be injected into the parasitic PNP of the PMOS device and induce latchup, which leads to catastrophic failure by shorting V_{DD} to ground [4.2]. This condition occurs:

$$\frac{dV_{DD}}{dt} \leq \frac{-V_{BE}}{\tau_{CLK|AVE}/2} \qquad \text{(EQ 4.9)}$$

where $\tau_{CLK|AVE}$ is the average clock period as V_{DD} varies from V_{out} to $V_{out} - V_{be}$. Since the clock period is longest at lowest voltage, this is evaluated as V_{DD} ranges from $V_{MIN} + V_{be}$ to V_{MIN}, where $V_{MIN} = V_T + 100\text{mV}$. For our 0.6μm process, the limit is -20 V/μs, which will increase with process technology.

If V_{DD} ramps up by more than V_{Tp} by the end of the evaluation state, and V_{out} drives a PMOS device, a false logic low may be registered, giving a functional error. This condition occurs:

$$\frac{dV_{DD}}{dt} \geq \frac{V_{Tp}}{\tau_{CLK|AVE}/2} \qquad \text{(EQ 4.10)}$$

evaluated for $\tau_{CLK|AVE}$ as V_{DD} varies from V_{MIN} to $V_{MIN} + V_{Tp}$. For our 0.6μm process, the limit is 24 V/μs, which will also increase with process technology because clock frequency improvement generally outpaces threshold voltage reduction.

These limits assume that the circuit is in the evaluation state for no longer than half the clock period. If the clock is gated, leaving the circuit in the evaluation state, these limits drop significantly. Hence, the clock should only be gated when the circuit is in the precharge state.

These limits may be increased to that of static CMOS logic using a small bleeder PMOS device, as shown in Figure 4.31. The left circuit can be used in logic styles without an output buffer (e.g. NP Domino), but has the penalty of static power dissipation. The right circuit is more preferable, as it eliminates static power dissipation, and only requires a single additional device in logic styles with an output buffer (e.g. Domino, CVSL). Since the bleeder device can be made quite small, there is insignificant degradation of performance due to the PMOS bleeder fighting the NMOS pull-down devices.

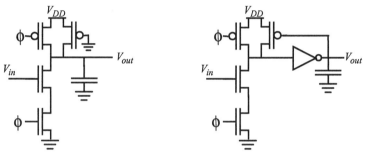

FIGURE 4.31 Two static latches with bleeder transistors.

4.5.2 Tri-state Busses

Tri-state busses that are not constantly driven for any given cycle suffer from the same two failure modes as seen in dynamic logic circuits due to their floating capacitance. The resulting dV_{DD}/dt can be much lower if the number of consecutive cycles in which the bus remains floating is unbounded. Tri-state busses can only be used if one of two design methods are followed.

The first method is to ensure by design that the bus will always be driven. This is done easily on a tri-state bus with only two drivers. The enable signal of one driver is simply inverted to create the enable signal for the other driver. However, this becomes expensive to ensure by design for a large number of drivers, N, which require routing N enable signals.

The second method is to use small cross-coupled inverters in order to continuously maintain state on the bus. This is more preferable to just a bleeder PMOS as it will also maintain a low voltage on the floating bus. Otherwise, leakage current may drive the bus high while it is floating for an indefinite number of cycles. The size of these inverters can be quite small, even for large busses. For our 0.6µm process, an inverter can readily tolerate a dV_{DD}/dt in excess of 75 V/µs with minimal impact on performance, and only a 10% increase in energy consumption.

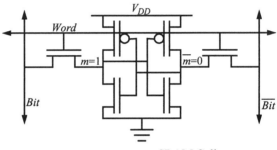

FIGURE 4.32 SRAM Cell.

4.5.3 SRAM

SRAM is an essential component of a processor. It is found in the processor's cache, translation look-aside buffer (TLB), and possibly in the register file(s), prefetch buffer, branch-target buffer, and write buffer. Since all these memories operate at the processor's clock speed, fast response time is critical, which demands the use of a sense-amp. The static and dynamic CMOS logic portions (e.g. address decoder, word-line driver, etc.) of the memory respond to a changing supply voltage similar to the ring oscillator, as desired. The sense-amp, however, must be carefully designed to scale in a similar fashion.

The basic SRAM cell is shown in Figure 4.32. *Bit* and $\overline{Bit}$ are precharged to the V_{DD} value at the end of the precharge cycle. While in the precharge state, both *Bit* and $\overline{Bit}$ will track any variations on V_{DD}. Once the *Word* signal has been activated to sense the cell, the precharge devices are disabled and *Bit* and $\overline{Bit}$ do not respond to V_{DD} variations. However, the current drawn through the pass device connecting $\overline{m}$ to $\overline{Bit}$

Design Constraints for Varying Voltage

will track V_{DD} since *Word* tracks V_{DD}, creating a voltage differential across *Bit* and $\overline{Bit}$ which will also track V_{DD}. So, to first order, the current drawn by the SRAM cell, which is a measure of its delay, will track V_{DD} similar to the delay of static CMOS logic.

If V_{DD} drops, m will drop, but since *Word* will also drop, there is no affect on *Bit* since the pass device is in the off state. When V_{DD} increases, m will increase, as will *Word*, but will have no affect until V_{DD} increases by V_{Tn}, which is required to turn on the pass device. When this occurs, this second-order effect will cause the voltage differential on *Bit* and $\overline{Bit}$ to increase faster because the pass device will begin charging up *Bit*, while $\overline{Bit}$ continues to be discharged. However, a dV_{DD}/dt in excess of 50 V/μs is required to induce this effect. Another second-order effect on the current drawn is that since *Bit* and $\overline{Bit}$ do not vary in the evaluation state with V_{DD}, the V_{ds} of the pass device remains constant, independent of any change on V_{DD}. However, this effect also requires large dV_{DD}/dt in excess of 50 V/μs to have any appreciable effect.

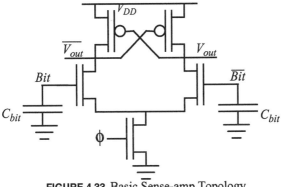

FIGURE 4.33 Basic Sense-amp Topology.

The basic sense-amp topology, shown in Figure 4.33, responds to the varying V_{DD} to first-order similar to static CMOS logic. Since the SRAM cell generates a voltage differential on *Bit* and $\overline{Bit}$ which scales with V_{DD}, the amount of time to generate the critical ΔBit to trip the sense-amp also scales. If the common-mode voltage between *Bit* and $\overline{Bit}$ were to scale with V_{DD} as it varies during the sense-amp evaluation state, then the delay of the sense-amp would scale much like static CMOS logic. However, this is not the case, and introduces the limiting second-order effect. As V_{DD} increases, the critical ΔBit to trip the sense-amp decreases, speeding the sense-amp up. As V_{DD} decreases, the critical ΔBit to trip the sense-amp increases, slowing the sense-amp down.

Circuit Design Methodology

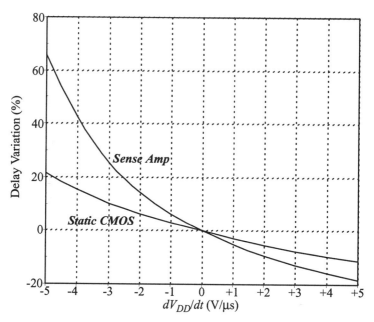

FIGURE 4.34 Sense-Amp Delay Variation with Varying Supply Voltage.

Figure 4.34 plots the relative delay variation of the sense-amp compared against the relative delay variation for static CMOS for different rates of change on V_{DD}. It demonstrates that the delay does shift to first order, but that for negative dV_{DD}/dt, the sense-amp slows down at a faster rate than static CMOS. For the prototype design, the sense-amp delay was approximately 25% of the cycle time. The critical path containing the sense-amp was designed with a delay margin of 10%, such that the maximum increase in relative delay of the sense-amp as compared to static CMOS that could be tolerated was 40%. This set the ultimate limit on how fast V_{DD} could vary in our 0.6μm process:

$$|dV_{DD}/dt| \leq 5V/\mu s \qquad \text{(EQ 4.11)}$$

This limit is proportional to the sense-amp delay, such that for improved process technology and faster cycle times, this limit will improve. What must be avoided are more complex sense-amps whose aim is to improve response time and/or lower energy consumption for a fixed V_{DD}, but fail for varying V_{DD}. One example is a charge-transfer sense-amp *[4.5][4.8]*.

4.5.4 Summary

As was demonstrated for the sense-amp, simpler circuit design ensures greater DVS compatibility. Many circuit design techniques developed for low power, such as the charge-transfer sense-amp and NMOS pass-gate logic, are not amenable to DVS. However, the potential energy efficiency improvement of DVS far outweighs the slight degradation in energy efficiency by not using these more energy-efficient circuit design techniques.

In addition, a design approach was assumed that requires that no signal should be floating for more than a half-cycle. But even with this approach, there are limits to dV_{DD}/dt, on the order of 20 V/μs for our 0.6μm process. Higher dV_{DD}/dt can be tolerated for dynamic circuits with the use of bleeder and feedback devices, but is not required since the sense-amp is the limiting factor. While the basic sense-amp's delay scales to first-order with dV_{DD}/dt, second-order effects limit $|dV_{DD}/dt|$ to only 5 V/μs in our 0.6μm process, a limit that will improve with scaling of the CMOS process technology.

References

[4.1]   T. Burd, *Low-Power CMOS Library Design Methodology*, M.S. Thesis, University of California, Berkeley, Document No. UCB/ERL M94/89, 1994.

[4.2]   N. Weste and K. Eshraghian, *Principles of CMOS VLSI Design*, Addison Wesley, Reading, MA, 1993.

[4.3]   D. Sylvester and K. Keutzer, "Getting to the bottom of deep submicron", *Proceedings of the International Conference on CAD*, 1998, pp. 203-211.

[4.4]   R. Ho, K. Mai, M. Horowitz, "Scaling Implications for CAD", *Proceedings of the IEEE International Conference for Computer-Aided Design*, Nov. 1999.

[4.5]   A. Burstein, *Speech Recognition for Portable Multimedia Terminals*, Ph.D. Thesis, University of California, Berkeley, Document No. UCB/ERL M97/14, 1997.

[4.6]   A. Stratakos, *High-Efficiency, Low-Voltage DC-DC Conversion for Portable Applications*, Ph.D. Thesis, University of California, Berkeley, 1998.

[4.7]   P. Gray, R. Meyer, *Analysis and Design of Analog Integrated Circuits*, Wiley, New York, 1993.

[4.8]   S. Kawashima, et. al., "A Charge-Transfer Amplifier and an Encoded-Bus Architecture for Low-Power SRAM's", *IEEE Journal of Solid State Circuits*, Vol. 33, No. 5, May 1998, pp. 793-9.

CHAPTER 5   *Energy Driven Design Flow*

The most critical aspect of energy-efficient design is to be energy conscious throughout the entire design flow. A typical design flow treats energy consumption as an afterthought, and is not thoroughly analyzed until the design has reached the transistor schematic stage. This is too late in the design process for radical modifications that can lead to a more energy-efficient implementation. Therefore, much as performance is analyzed at the initial high-level specification of the design, so must energy consumption be analyzed, as well. The primary goal of this design flow is to evaluate energy consumption early on so that the largest energy reductions can be attained.

In today's complex chip designs, a majority of the design cycle is spent on validating a design for proper functionality, and verifying its layout implementation. Implementing DVS exacerbates the verification problem by requiring multiple operating conditions to be analyzed. Another goal of this design flow is to automate design validation and verification so that the bulk of the design effort could remain focused on the design implementation and optimization for energy efficiency.

The first section presents an overview of the design flow. Subsequent sections explore in more detail those parts of the design flow that were developed to aid in the design and verification of a DVS microprocessor system, though most of this design flow is equally applicable to general energy-efficient digital design.

5.1 Overview

The basic design flow from system specification to final chip layout is presented in Figure 5.1. The flow refines the design through five discrete phases, each of which has its own set of design tools, and optimization goals. Through each refinement phase, the design is verified against the previous phase's implementation to insure proper functionality. Through transitive equivalence, the layout can be verified to operate as specified by the initial cycle-level simulator.

Implementing an existing instruction set architecture (in this particular case the ARM V4 ISA *[5.1]*) provides the key benefit of having proven compilers and assemblers available. The design verification process is bootstrapped by the ability to swiftly write self-validating C code which can be compiled down to machine code and executed on the simulator. This allows the simulation of real test code early on with abstract design models.

An advantage of implementing the ARM V4 ISA was the availability of a simulator (from ARM Ltd.) for the processor core *[5.2]*; once an abstract memory/IO module was written, the simulator could begin booting the operating system and executing benchmark programs. This provided the ability for high-level performance and energy estimation so that the design could be optimized as the simulator evolved into a cycle-accurate specification. Performance tuning at this stage is common in microprocessor design, but energy estimation is typically done as an afterthought. By incorporating energy estimation into the simulator, high-level energy optimization significantly improved the design's energy efficiency as will be discussed in Section 5.2.

Once the performance and energy optimized cycle-accurate simulator was developed, the design progressed to the VHDL behavioral design phase. Since the design is still at a behavioral-level specification, no significant optimizations are possible at this level. Since the specification language has changed from C to VHDL, it is critical at this juncture that the two models behave exactly the same, as discussed in Section 5.5.1. The design was not initially specified in VHDL because C simulation is several orders of magnitude faster, which enabled a rapid design turn-around during the initial specification phase.

As the VHDL description transitions from behavioral to structural models, additional performance and energy optimization occurs. By annotating the structural VHDL model with delays, critical paths can be identified and reduced. At this microarchitecture level, better energy estimates are available for the individual blocks, and the sim-

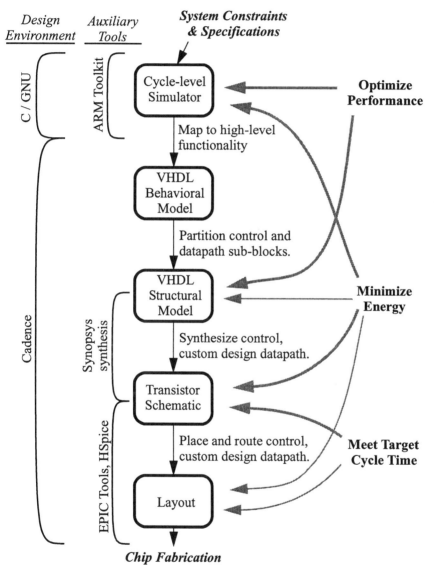

FIGURE 5.1 General Design Flow from Specification to Layout.

ulator is updated accordingly to provide better energy estimates. Dominant energy consuming blocks can be identified and optimized. By the end of this phase, remaining critical paths that cannot be removed via microarchitecture changes dictate the achievable cycle time. At this point, the performance optimization transforms to meeting the targeted cycle time.

Another significant level of energy optimization occurs during the transistor design utilizing the various energy-efficient design techniques in Section 4.1. During this phase, PowerMill provides accurate energy estimates which aid the designer to further reduce energy consumption. Accurate timing analysis (Section 5.6) ensures the target cycle time can be met. Simple schematic redesigns are done when the target cycle time cannot be met, as indicated by long path lengths, which is much less costly than waiting for timing results on the extracted layout.

During the final layout phase, energy again can be minimized through intelligent layout, but to a much less degree than previous phases. Through LVS, the layout can be matched back to the schematic to verify a correct design. Timing analyses on an extracted layout netlist can flag additional critical paths that need to be fixed which were not flagged during the schematic design phase.

5.1.1 Energy Budgeting

To most effectively optimize energy consumption, energy reduction in total system energy must be measured, and not just localized energy reduction. For example, if a particular design change reduces the energy consumption of the write buffer 50% while degrading system performance by only 10%, it may appear to be a desirable optimization. If, however, the write buffer only consumes 10% of the overall energy, then a 5% reduction in overall system energy consumption does not justify a 10% performance reduction. Thus, it is imperative to always evaluate energy and performance at the system level for potential design optimizations, and not in isolation.

Section 5.2 describes in further detail a methodology for high-level system energy estimation. This provides an estimated breakdown of the system energy consumption long before the design is taken to structural and physical implementation. This breakdown provides crucial guidance on what blocks require careful design to minimize energy, and what blocks consume negligible energy and can be rapidly designed through synthesis. This breakdown is updated and maintained throughout the design process to track where the focus should be for minimizing system energy consumption.

5.1.2 Verification Overview

Verification checks are performed at each level of the design phase, as shown in Figure 5.2. At the C & VHDL behavioral levels, the checks are strictly for functionality. Test code is created using the methodology described in Section 5.5.2, which verifies that the behavioral models match the specification for the ISA and IO expected by the programmer. Scripts automatically generate test vectors for use at the structural and transistor level design phases.

Very simple timing analysis is performed at the structural model by simulating timing-annotated VHDL models. Excessively long critical paths can be identified at this stage, ensuring an implementation is possible with the target cycle time derived at the end of this design phase. Comprehensive timing analyses are performed at the schematic and layout design phases, as described in Section 5.6.

Transistor sizing checks are done at the schematic design phase to catch grossly under/oversized clock buffers and bus drivers, based upon capacitance estimates from the schematic netlist. Once the design transitions to the layout phase, final sizing checks are performed on the extracted netlist to guarantee valid clocking operation (Section 5.3) and to minimize short-circuit current (Section 4.1.2).

Finally, the layout requires standard additional checks to verify a proper implementation (LVS, DRC, etc.). Through the use of scripts, all these tasks can be automated.

5.2 High-level Energy Estimation

High-level energy estimation for architectural exploration has been successfully demonstrated within a VHDL simulation environment *[5.3]*. While this approach is suitable for dedicated DSP architectures it is not a suitable approach for general-purpose processor systems which require several orders of magnitude more simulation cycles to properly characterize the system. A reasonable benchmark suite requires billions of simulated cycles; even on the fastest VHDL simulator, the suite would take at least a day to complete a single pass.

To reduce simulation time, processor behaviorial models can be written in C, which can speed up simulation time one to two orders of magnitude, and then used to provide extensive performance statistics. However, they can also be augmented to pro-

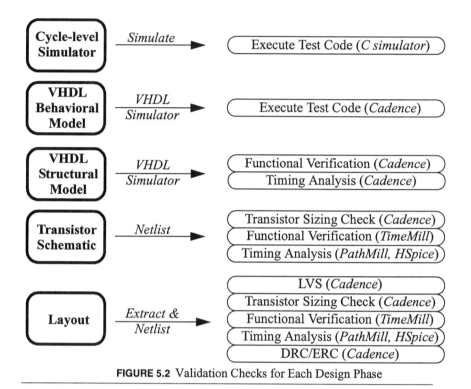

FIGURE 5.2 Validation Checks for Each Design Phase

vide energy consumption statistics as well, by applying the same black-box capacitance modeling used in the VHDL simulator to the C simulator.

This enables the high-level architectural exploration methodology shown in Figure 5.3 which has modified a standard ARM v4 processor core simulator to provide both performance and energy statistics. Initial hardware capacitance estimates are added to the simulator, which then executes the benchmark suite to provide statistics for identifying dominant energy-consuming blocks and performance bottlenecks. New optimizations are proposed to improve energy efficiency, and implemented in the C simulator, so that the improvement in energy efficiency of these design optimizations can be quantified after resimulating the benchmark suite. An example simulator output report file is shown in Figure 5.4.

High-level Energy Estimation

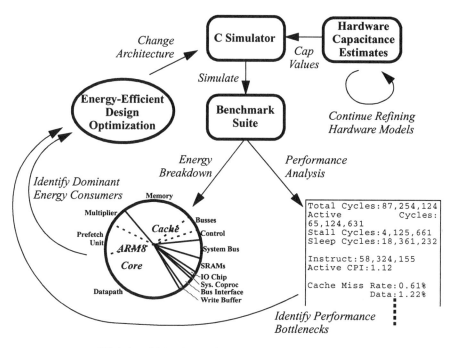

FIGURE 5.3 High-level Simulation for Performance & Energy Estimation

Energy Estimation & Execution Statistics

The base simulator is annotated with mechanisms for gathering energy estimates and execution statistics. Individual mechanisms can be selectively enabled or disabled using compile-time flags, allowing specific attributes to be analyzed without adversely affecting performance. The simulator report contains the following general sections:

- **Block-level energy estimation**: At the lowest level of energy estimation, each usage of a block, such as the ALU, consumes a fixed amount of energy per-use based on average-case execution. Specific instructions, such as add-immediate, are responsible for calling the appropriate energy accounting routines for the blocks that they use. More complicated instructions, such as load-multiple (LDM), tend to invoke more block and therefore consume more energy. Fine-grained optimizations, such as aligning memory accesses to reduce the number of bit transitions will have no effect on this level of estimation.

Energy Driven Design Flow

Energy Driven Design Flow

```
Sim time:     155.0ms
Run time:     118.0s
Simulation speed 1/761th

Energy Estim. for ARMulator @ level 1
--------------------------------------
Accesses to MUL Sim Block:       8580
Accesses to ADM Sim Block:        280          Debug Info to Verify
Accesses to PSR Sim Block:       5963          Correct Operation
Accesses to ALU Sim Block:    3393973
Accesses to IMM Sim Block:    1627554
...
Block   Accesses  Energy (%)  Sw. Cap (%)
-----   --------  ----------  -----------
S/ALU   5663245    9.446 %      8.792 %
RegBk   6768302    0.001 %      0.001 %       Block Energy Consumption
Mult       8580    0.048 %      0.044 %
PSR      241694    0.134 %      0.125 %
...
Bus  Accesses   Energy     Sw. Cap    Toggle
---  --------   -------    -------    -------
A    5714811    5.224 %    4.862 %    25.687 %
B    3570615    2.591 %    2.411 %    27.185 %
RES  4433936    4.592 %    4.274 %    29.106 %   Bus Energy Consumption
CDT      156    0.000 %    0.000 %     0.401 %
IO      9662    0.048 %    0.044 %    11.071 %
...
Delta Energy = 0.0195862 Joules
Total Elapsed Time = 0.268321 sec
Delta Elapsed Time = 0.155443 sec           System
Delta Instructions = 5915641                Summary
  Removed Branches = 0
Delta Exec. Cycles = 7551141
               CPI = 1.27647
Energy / Instruct. = 3.31092 nJ
     Average Power = 126.003 mW

Instruction count report:
Data processing : 75.56%
    Set ccodes  :  5.32%
    op1 not used: 28.66%
    no result   :  4.25%                    Instruction
PSR Op          :  0.06%                    Profiling
Swap            :  0.01%
Multiply        :  0.15%
...
5915k instructions total profiled.

Cache/Memory report:                        Configuration
Unified: 16kb/8w-lines/4-way/Lru             Info
FCLK: 50MHz   MCLK: 25MHz    WB: 8g   NCW_NWB

       __hits__   __miss__  _rate_
Data   1239206       7781    0.63%
Inst   3478202      17461    0.50%          Activity
Totl   4717408      25242    0.54%          Profiling

active portion: 1.000000e+00
actv : 4743884  sleep:       0  nmreq:  2777779
cwait:  220987  iwait:   29478  total: 77721282
wbacc:   20966  wbful:     403
...
```

FIGURE 5.4 Portion of a Simulator Report File.

High-level Energy Estimation

- **Bit-level energy estimation**: Bit-level energy estimation monitors bit-level activity to better determine the energy consumed. For example, a change from 0x100 to 0x101 on the address bus triggers one bit-flip, while 0x100 to 0x0ff triggers nine bit-flips and will therefore consume nine times as much energy. This level of estimation requires more storage and computation than the block-level estimation, resulting in slower simulations. Furthermore, complicated components, such as the ALU or shifter, do not necessarily follow a liner relationship between bit flips and energy consumed, reducing the accuracy of bit-level estimation.
- **Instruction usage summary**: Basic statistics on the kind of instructions executed are kept to profile the instruction mix for the different applications.
- **Processor/memory cycle counts**: Processor and memory bus cycles are counted, allowing a better understanding of the behavior of the entire system. For the processor, there are *instruction fetch*, *memory access*, *no request*, and *cache stall*. For the external memory bus, there are *idle*, *address*, and *data* cycles.
- **Cache statistics**: Statistics on the cache subsystem, such as miss rate, stall cycles, double-word hit rate, and tag-checks are counted allowing easy analysis of different cache geometries. Section 3.3 covers the analysis of different cache geometry trade-offs in detail.

Energy estimates were calibrated using early versions of the processor layout, which were then optimized using the estimates yielding new calibrations in a cyclic design process. The estimates are used in several different ways to improve the prototype processor system. The aggregate energy for a complete run of a particular benchmark is used to gauge the effectiveness of a specific voltage scaling algorithm: changing the processor speed at run-time alters the voltage the simulator thinks it is running at, which in turn effects the energy estimates. The block-level energy breakdown is used to direct processor optimization: for example, it is not worth the design effort to optimize the multiplier since it consumes less that 1% of the total energy. Similarly, instruction usage statistics are used to make better design decisions in general, and were often implemented to answer a specific question (such as the frequency of load-multiple instructions).

All design optimizations were parameterized where possible (e.g. cache-line length, write-buffer size, etc.), so that previous configurations could be re-simulated via command line flags. This was crucial since the hardware capacitance models were refined as the design progressed. Once the simulator incorporated the new capacitance models, all previous configurations could be resimulated to find any design optimization points that have shifted due to the newly refined models.

5.2.1 Capacitance Models

Energy consumption is one of the desired outputs from the simulator. To accommodate the variable V_{DD} in DVS, the underlying simulator models are based upon capacitance. During simulation, the capacitance is multiplied by V_{DD}^2, which varies depending upon the current clock frequency setting. There are varying approaches to capacitance estimation, with the simulator complexity increasing with estimation accuracy. The three simplest approaches are:

White-noise Approach (0th order): Every time a block or bus is accessed, a counter associated with that block is incremented by the average capacitance switched on that block/bus by V_{DD}^2. The switched capacitance values assume random signals on the blocks so that a single value can be used.

Data Correlation Approach (1st order): This approach uses simulator state variables associated with each block/bus's input ports to hold the previous state. The current and previous input values are XORed to count the number of transitioning bits, which are then multiplied by a capacitance per bit value. For a bus, this capacitance is just a sum of the estimated interconnect capacitance and fanout gate capacitance. For circuit blocks, an empirical estimate can be made from circuit simulation data. This works well on the majority of blocks that are bit-wise symmetrical, such as memory, register files, shifters, muxes, etc. Non-symmetrical structures that have inter-bit data correlation, such as an adder, will give an estimate with much greater variance because the capacitance is not only a function of how many bits transition, but the location of the transitioning bits as well.

Data/Instruction Correlation Approach (2nd Order): This approach builds on the previous approach by also keeping track of previous instruction(s) with simulator state variables which can be used to model the inter-instruction dependencies. This better accounts for energy consumption in various muxes and state latches which are used to route the data flow of the processor from instruction to instruction.

Previous work has shown that the white-noise approach does an inaccurate job of energy estimation, so to provide a reasonable (first order) estimate, data correlation must be accounted for. The second order approach will improve accuracy, but is not necessary to provide useful energy estimates.

5.2.2 Implementation

It is critical that the energy estimation code be efficiently implemented so that the speed of the C simulator is not significantly degraded. Thus, the counters and state registers required should utilize native integer word types on the host simulation machine. In addition, all functions should be compact and inlined to remove the unnecessary overhead of function calls.

A typical processor simulator consists of a large case statement representing the vari-

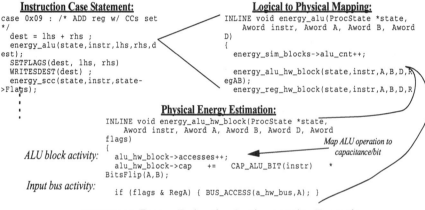

FIGURE 5.5 Energy Estimation Implementation Example.

ous instructions, or classes of instructions. As such, the energy estimation function calls required to annotate the simulator must be organized along instruction boundaries, and not physical organization. A second level of functions calls are used to map the logical system organization to an orthogonal microarchitecture specification, as demonstrated with the simple code example in Figure 5.5.

Fully annotating the simulator requires placing these logical energy function calls wherever the simulator changes the system state (e.g. *dest = lhs + rhs*). Using a simple system microarchitecture specification (Figure 5.6), these logical functions are then mapped to physical function calls.

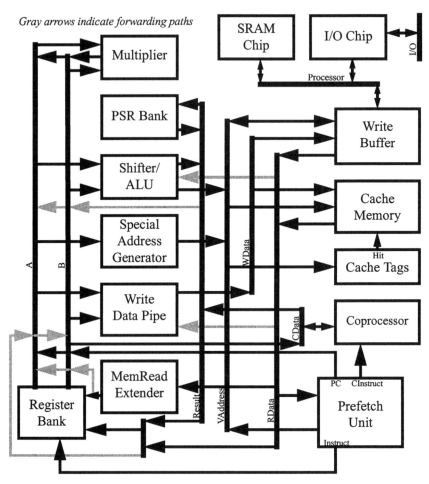

FIGURE 5.6 Microarchitecture Model for High-level Energy Estimation.

5.3 Clocking Methodology

A well-defined clocking strategy is required for any sub-micron digital integrated circuit. The goal of this strategy is to limit the clock skew between any two latches. The secondary reason is to prevent performance degradation. If the delay on a critical path is more than targeted, due to clock skew, then the overall cycle time needs to be increased for proper functionality.

Maintaining DVS compatibility adds further constraints to the clocking methodology. The ratio of clock skew to cycle time should remain fixed over the range of operating voltage. While the delay through the clock buffers will scale, the RC delay on the interconnect does not. Hence, the critical design corner to meet is at high voltage, fast process, and low temperature.

5.3.1 Latch Design

The behavioral model for the processor core that was used as the design starting point assumed a latch-based design *[5.6]*, resulting in a latch rather than a flip-flop being the basic state element of the design.

A fully static latch approach was taken in spite of it not being the most energy-efficient. The compensating features are that it is the most robust *[5.5][5.7]* and well suited for use with aggressive clock gating (Section 4.1.3) since the clock can be held either high or low indefinitely without inducing a logical error. In contrast, a gated dynamic latch can have its internal state flipped via leakage currents and DVS voltage changes *[5.4][5.7]*. Additionally, static latches allow the system implementation to be stepped phase by phase, which is a tremendous advantage when debugging the hardware design.

Intuitively, it would seem that a single-phase latch is preferable over the traditional two-phase, cross-coupled latch because one less clock wire is needed. On the other hand, the natural design of a single-phase latch is dynamic. To make the latch static requires a more complex latch design, and effectively eliminates the advantage.

The basic latch design is shown in Figure 5.7, in which the feedback device is an inverter. Another possible implementation is to use a transmission gate between nodes *out* and *x*, but this can lead to a race condition on back-to-back latches if there is overlap on *clk* and *clkb*. The feedback device is clocked to both guarantee latch operation and speed up the latch. While the latch is transparent, there is no contention on the forward path, which greatly reduces the setup time, and allows any input signal to successfully change the value on node *x*. This does come at a penalty of an additional 15% in clock net capacitance for a datapath latch. The penalty will be less for a standard cell latch, where the interconnect capacitance is greater, thereby reducing the contribution from the feedback device.

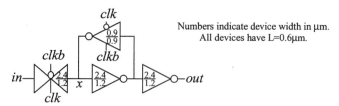

FIGURE 5.7 Basic Two-phase Latch Design.

Since the behavioral model constrained the basic state element to be a latch, the fully-static approach seemed to best balance energy efficiency and design robustness. Had the model assumed a flip-flop based design, then a pseudo-static version [5.4] of the TPSC register [5.8] would yield a more energy-efficient and robust design. This solution requires only one clock wire, yields equivalent register setup & hold times, and allows clock gating if the clock is held low.

5.3.2 Clock Architecture

The basic clock-gating strategy is split-level. There is a local enable signal for each n-bit latch, or set of latches. This signal is active-high, and is used to selectively generate a local clock pulse. There are also three global halt signals, which are used to halt large sections of the microprocessor: the processor core, the cache subsystem, and the bus interface. Rather than actively suppressing clock pulses, which will change the processor state, it simply gates out the low-phase of the global clock signal, as demonstrated in Figure 5.8, maintaining exact processor state.

This halting mechanism allows the processor memory interface to be designed to expect the cache subsystem to return a word within one cycle of the request. If the cache subsystem needs to go out to main memory, it simply halts the processor core until it has the desired word available, and eliminates any unnecessary switching activity in the core. Likewise the bus interface can stall the cache subsystem when it is retrieving words, and the external I/O system can stall the bus interface when a high-latency I/O memory request is made.

The natural implementation for this would be a three-level clock hierarchy, divided by the global (halt) clock drivers, and the local (enable) clock drivers. However, the

Clocking Methodology

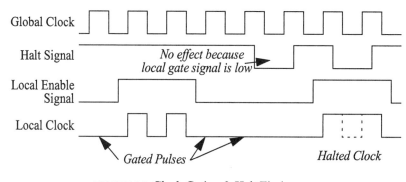

FIGURE 5.8 Clock Gating & Halt Timing.

skew can be better controlled in a two-level clock hierarchy by removing the skew contribution from the global clock driver, although this comes at the expense of increased energy consumption.

In the two-level hierarchy, the total capacitance on the global top-level clock net is 10pF, which is just 3% of the nominal effective capacitance switched per cycle. Another consideration is clock power when the processor is in the sleep mode. However, if the processor is put into a low-speed mode before entering sleep, the idle clock power dissipation is only 72μW, which is only 9% of the total system sleep mode power. Hence, it was decided a worthwhile trade-off to implement a two-level clock hierarchy which provided a more controlled clock skew. In addition, it made the timing analysis simpler, because both the local enable and global halt signals are terminated at the same gate.

Clock Drivers

One approach to generating the two-phase clock signals is to locally invert within the latch cell. However, this has two significant drawbacks. First, the additional gate gets toggled every clock cycle, which adds more capacitance to the global clock net as compared to the interconnect capacitance added by running a second clock wire. Second, the skew between the two clock signals is now a full gate delay, which is not tolerable as will be demonstrated in the clock skew analysis later on. To reduce this skew, the non-inverted clock signal can be buffered by two inverters, but since these toggle every cycle, further capacitance is added to the global clock net. Hence, it is preferable to generate the two-phase signal once per n-bit latch. For the datapath registers, which dominate the overall register count, this requires one clock driver per 32-bit latch, and can be designed to have well-controlled skew.

Energy Driven Design Flow

Before each n-bit latch is a clock driver which performs single to differential polarity conversion, as shown in Figure 5.9. A phase-1 clock driver generates a high pulse on

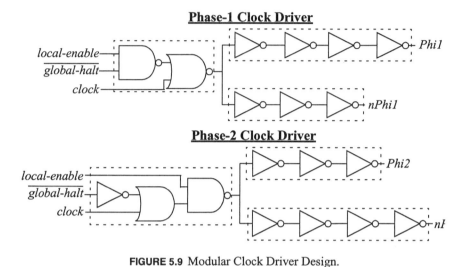

FIGURE 5.9 Modular Clock Driver Design.

the *Phi1* signal while *clock* is low, and the *local-enable* and *global-halt* (active low) signals are high. Similarly, the phase-2 clock driver generates a high pulse on *Phi2* when *clock* is high, and the *local-enable* and *global-halt* signals are high. The *local-enable* signal must only transition in the phase opposite of the clock driver phase (e.g. a phase-1 clock driver's *local-enable* signal can only change while *clock* is high, such that it must be generated by a phase-2 latch, or derived only from signals output by phase-2 latches). The *global-halt* signal, which drives clock drivers of both polarities, can only transition while *clock* is high.

The maximum skew between any two clock drivers was determined in Section 5.3.4 to be less than one-half of a gate delay. To maintain controlled skew, the inverter chains need to be carefully sized for the output load, which can vary from less than 50fF to 1pF. Close to 150 clock drivers where used in the design. Rather than design a new clock driver each time one was required, and insure that it met the skew constraint over process and voltage, a modular clock driver design was developed.

This modular approach consists of 2 gate cells, one for each phase, and 24 each of inverting, and non-inverting driver chains. The modular cells in Figure 5.9 are denoted by dotted lines. The input capacitance to the driver chain was kept constant so that the same gate cell could be used for both small and large drivers, and not require re-tuning. Since the input inverter is a fixed size, and the output inverter size is solely dictated by the load being driven, the non-inverting clock path required four inverters, rather than just two, to properly tune the driver delay. However, the extra two inverters consume an insignificant amount of energy on all but the smallest clock drivers. In addition, the layout took a modular approach, with the gate cell in the middle and a driver chain cell abutting either side of it. The most significant benefit of this modular approach was that the 50 unique cells could be designed and verified independent of the final chip design.

Table 5.1 summarizes the delay and skew of the entire clock driver suite at three different voltages and across process variation. The values have been normalized to the clock cycle time, t_{CLK}, to demonstrate that both the mean delay, and the skew track extremely well over V_{DD}.

TABLE 5.1 Clock Driver Delay and Skew (0.6μm Process).

V_{DD}	Process Corner	Delay (ps / % of t_{CLK})			Skew
		Mean	Max.	Min.	(ps / % of t_{CLK})
4.1	Fast	555 / 7.9%	590 / 8.4%	510 / 7.3%	80 / 1.1%
	Nominal	709 / 7.9%	755 / 8.4%	650 / 7.2%	105 / 1.2%
	Slow	897 / 7.8%	960 / 8.3%	826 / 7.2%	134 / 1.2%
3.3	Fast	648 / 8.6%	687 / 9.2%	601 / 8.0%	86 / 1.1%
	Nominal	835 / 8.4%	885 / 8.9%	770 / 7.7%	115 / 1.2%
	Slow	1070 / 8.6%	1140 / 9.1%	989 / 7.9%	151 / 1.2%
1.1	Fast	5510 / 7.3%	5970 / 8.0%	5100 / 6.8%	870 / 1.2%
	Nominal	8870 / 7.4%	9640 / 8.0%	8080 / 6.7%	1560 / 1.3%
	Slow	17300 / 4.5%	19300 / 5.1%	15000 / 3.9%	4300 / 1.1%

When a new clock driver was designed into the chip implementation, the capacitance on both *Phi* and *nPhi* were estimated from the schematic. These values were used to index into Table 5.2 to determine the driver size for the two signals. If the desired clock driver cell currently existed, it would be used, and if not, then a new clock driver can be rapidly constructed. Once the layout was complete, the design was

Energy Driven Design Flow

extracted to measure the final signal capacitance. If the sizings were found to be wrong for the extracted capacitance, a new clock driver could easily be inserted.

TABLE 5.2 Clock Driver Sizing (not including diffusion capacitance)

Driver	Min Load (fF)	Max Load (fF)	Driver	Min Load (fF)	Max Load (fF)
1x	37.5	62.5	13x	450	525
2x	50	87.5	14x	487.5	562.5
3x	75	125	15x	525	600
4x	112.5	187.5	16x	562.5	637.5
5x	150	225	17x	600	675
6x	187.5	262.5	18x	637.5	712.5
7x	225	300	19x	675	750
8x	262.5	337.5	20x	712.5	787.5
9x	300	375	21x	750	825
10x	337.5	412.5	22x	787.5	862.5
11x	375	450	23x	825	900
12x	412.5	487.5	24x	862.5	937.5

It would seem that only 48 unique clock drivers are needed, assuming the output signals having equal driver strength (1-24x for phase-1, and 1-24x for phase-2). However, the majority of clock drivers had differently sized driver chains due to varying interconnect capacitance. Hence, this modular design approach worked quite well by allowing the rapid assembly of any arbitrarily-sized clock driver.

5.3.3 Bounds on Allowable Skew

The worst-case race condition between back-to-back latches sets the maximum allowable clock skew *[5.5]*. Any logic in between the latches makes the circuit more robust against clock skew. There are two different cases to consider. The same-latch case occurs when the two latches are clocked by the same two clock signals, such as in a flip-flop, and there is skew between *Phi* and *nPhi*. The unrelated-latch case occurs when there are two sets of clock signals, *PhiA/nPhiA* and *PhiB/nPhiB*, and the skew occurs between the two sets. For each of these cases, the maximum allowable clock skew, t_{MAX}, was found so that the two latches remain immune to clock race, as shown in Figure 5.10.

Clocking Methodology

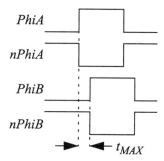

FIGURE 5.10 Maximum Allowable Clock Skew Measurement for Race Immunity.

Spice simulations were run to measure t_{MAX} over voltage and process, as given in Table 5.3. As previously mentioned, the critical corner to design for is high voltage and fast process because interconnect RC delay will be most significant there. The other corners were measured to ensure that the clock drivers scaled properly over voltage. These results are used in Section 5.3.5 to verify that there will be no fatal errors in the chip implementation due to race conditions.

TABLE 5.3 Maximum Allowable Clock Skew (in ps).

V_{DD}	Process Corner	t_{MAX} (unrelated-latch)	t_{MAX} (same-latch)
4.1	Fast	320	300
	Nominal	410	390
	Slow	530	490
3.3	Fast	380	380
	Nominal	500	480
	Slow	640	610
1.1	Fast	3700	3800
	Nominal	6000	6100
	Slow	10500	15000

Energy Driven Design Flow

Energy Driven Design Flow

5.3.4 Sources of Skew

There are several sources of clock skew, and each component was carefully analyzed and quantified. A total maximum clock skew can be calculated by summing up the individual components. While this is a conservative estimate because it assumes no correlation in skew, it has a high confidence level because each component has been completely analyzed and accounted for. The location of these sources are shown in Figure 5.11.

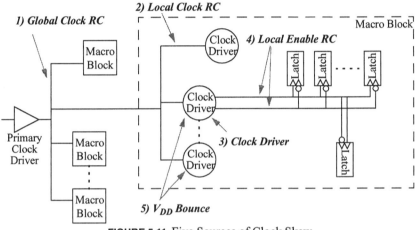

FIGURE 5.11 Five Sources of Clock Skew.

The initial analysis assumed that the two dominant sources of skew were the global clock RC and the clock driver skew. Each component was given a delay budget equal to 25% of the maximum allowable skew. This gave 100% headroom for margin of error and for additional skew components. Upon completion of the design, the various components were re-analyzed to verify the non-existence of race conditions.

Global Clock Wiring

This component is due to the voltage-independent RC delay on the global clock wire. It is measured as the skew between the clock input signals each macro block sees, where each macro block was modeled as a lumped capacitance. To make sure the delay budget of 25% of the maximum allowable skew could be met, an initial chip

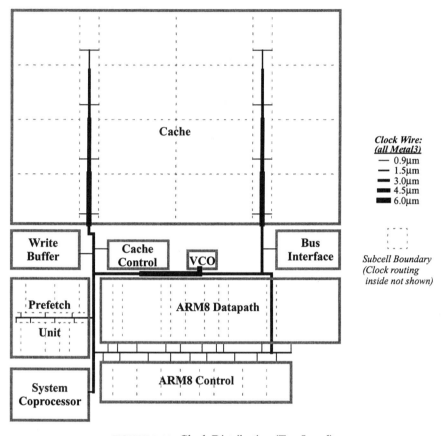

FIGURE 5.12 Clock Distribution (Top Level).

floorplan was analyzed. A very simplistic model demonstrated that this target could be met, and only require 3-4x larger than minimum-size wiring.

At the top level, widening the clock wire was very beneficial because the fractional capacitance of the clock interconnect was small compared to all the clock drivers' gate capacitance. Hence, widening the wire almost linearly reduced the RC delay.

During the course of the implementation, a regular 20μm-wide clock routing channel was created to allow for post-layout wire widening, and to eliminate interline capacitance. The clock distribution attempted to model an H-tree distribution network. After the initial chip layout was completed, a parameterized model of the global clock wire was created that modeled the clock wire as a distributed RC network with lumped capacitances for each macro block. The widths of the various wire segments were parameterized so that the RC delay could be tuned without having to extract the layout for each iteration.

The tuned top-level clock routing for the prototype processor is shown in Figure 5.12. The routing used third-level metal (*Metal3*) almost exclusively, due to its low resistivity, except where the clock routing intersected the top-level power grid network. The final simulation yielded a maximum RC delay between any two macro blocks of 81ps. The maximum delay between any two macro blocks within the processor core was just 31ps. The widest wire segment was 6.0μm, with an average wire segment of approximately 2.4μm.

Local Clock Wiring

This component is the RC delay on the clock signal within each macro block, measured as the largest skew seen by any two clock drivers. This component was designed to be less than 15ps for the largest block, the ALU control block, providing 25% margin.

This RC delay is negligible because the longest wire is no more than 1mm, as dictated by the bounded macro block size. Care was taken in routing this clock signal; *Metal3* was predominantly used, with the minimum number of vias. The 15ps estimate above was for an initial place and route of the ALU control block with no special attention paid to the signal routing, increasing the conservatism of the estimate.

Clock Driver Skew

The drivers are a dominant source of clock skew, and were designed to have a maximum skew across all possible drivers no more than 25% of the maximum allowable skew. Achieving this goal required careful design and sizing of the clock drivers. The largest driver was implemented first to ensure it could be designed, and then smaller drivers were designed to be within the delay budget. For the smallest drivers, additional internal capacitance had to be added to meet the targeted delay variation.

The 25% rule was most critical at high voltage, where the RC delay is also significant. At the low voltage corner, the allowable margin was increased to 50% since the RC delay becomes negligible. The maximum delay variations for the entire suite of clock drivers is given in Section 5.3.5, and meets the targeted specification.

Clocking Methodology

Local Enable Wiring

This component is the RC delay on the *Phi* and *nPhi* latch-enable signals. The worst case for this arises in the datapath, in which the enable signal must traverse 950μm across a 32-bit latch. This was measured to be 20ps, and relatively negligible. To provide a margin for error, this component was set to 30ps.

In the routed control blocks, particular care was taken in routing the enable signals. The placement was optimized to group latches around similar enable signals. Additionally, these routes were given the highest priority, along with the local clock signal, to optimize their routing. In the final layout of the synthesized blocks, due to the latch clustering, no enable signal ran longer than 1mm validating the 30ps estimate.

Enable Rise/Fall Time

Because the enable signals have a finite rise/fall time, there is a finite time that the latches can remain open even if the signals' skew, as measured between 50% points, is zero. Shown in Figure 5.13 are Spice simulations results which demonstrate that the maximum allowable skew actually increases monotonically with rise/fall time, despite the increasing overlap of the enable signals. Hence, the maximum allowable skew, as measured with step edges in Section 5.3.3, is a conservative estimate of the skew.

This component, if anything, would contribute a negative number to the maximum skew calculation. However, it is simply ignored, increasing the conservatism of the maximum allowable skew calculation.

V_{DD} Variation Skew

The power distribution network (Section 5.4) was designed so that no more than a 10% critical-path delay variation occurs with V_{DD} variation. V_{DD} variations only affect the clock driver delay, which has a mean delay just under 10% of the target critical-path delay. Thus, the maximum skew from V_{DD} variation is just a product of these two percentages, or just 1%. This component is voltage dependent, and will vary with V_{DD} similar to clock driver skew.

5.3.5 No-race Verification

To verify that race conditions cannot exist, the individual skew components were summed up and compared against the maximum allowable skew. Due to the initial

Energy Driven Design Flow

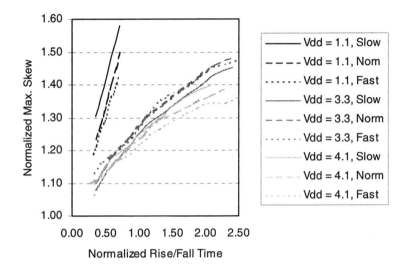

FIGURE 5.13 Skew as a Function of Enable Rise/Fall Time.

delay budgeting on the clock drivers, and the global clock wire analysis, the final implementation met the targeted skew with some margin.

TABLE 5.4 Total Clock Skew Summary (all times are in ps): Unrelated Latches

V_{DD}	Process Corner	Global Clock RC	Local Clock RC	Clock Driver	Local Enable RC	V_{DD} Bounce	Total Skew	Maximum Allowable Skew
4.1	Fast	80	20	80	30	70	280	320
	Nom			105		90	325	410
	Slow			134		115	380	530
3.3	Fast			86		75	290	380
	Nom			115		100	345	500
	Slow			150		125	405	640
1.1	Fast			870		750	1750	3700
	Nom			1560		1200	2890	6000
	Slow			4300		3800	8230	10500

Table 5.4 gives the comparison for the unrelated-latch case. As expected, the smallest skew headroom (Maximum Allowable Skew - Total Skew) occurs at high voltage and fast process, and is only 13% of the allowable skew which is adequate due to the conservatism built into the estimate. The same-latch case was much easier to meet, as shown in Table 5.5 . Since in this case the two latches share the same *Phi* and *nPhi* signals, global and local clock RC are made irrelevant because the latches share the same clock driver.t

TABLE 5.5 Total Clock Skew Summary (all times are in ps): Same Latch

V_{DD}	Proc.	Clock Driver	Local Enable RC	V_{DD} Bounce	Total Skew	Max Allowable Skew
4.1	Fast	80		70	180	300
	Nom	105		90	225	390
	Slow	134		115	280	490
3.3	Fast	86		75	190	380
	Nom	115	All 30	100	245	480
	Slow	150		125	335	610
1.1	Fast	870		750	1650	3800
	Nom	1560		1200	2790	6100
	Slow	4300		3800	8130	15000

Thus, by design, the chip implementation should be free from any race conditions, across the entire voltage and process range. The only precondition is that the latch contains two inverters. The number of inverters were reduced in a few latches due to critical path constraints, but in these cases it was ensured that there was no latch immediately following the faster latch.

5.4 Power Distribution Methodology

Typical low-power chip designs have very relaxed constraints on the power distribution network due to low DC and peak supply currents. While a DVS processor generally has a low DC supply current, the peak supply current can be quite high when it is operating at maximum clock frequency and supply voltage, thereby placing tight constraints on the power distribution network, both at the chip level and at the board level. Thus, power distribution requires careful design consideration in a DVS system similar to any high performance, and high current, chip design.

5.4.1 On-chip Supply Variation

The on-chip voltage supply will vary due to inductive and/or resistive voltage drops across the chip's power distribution network, and across the pins and bonding wires of the chip's packaging. Global supply variations, which occur uniformly across the chip, are essentially no different than changing the external supply voltage in a DVS system. Thus, the DVS chips are relatively immune to global on-chip supply variations. However, the problems arising from local supply variations, which occur within a limited area of the chip, are the reduction of signal noise margin and timing violations, both of which can induce functional failure.

Noise Margin Reduction

Static CMOS circuits and most dynamic logic circuits (e.g. Domino, NORA, DCVSL, etc.) are very robust against noise margin reduction, since their signal swing is the full value of V_{DD}, and have a noise margin of at least V_T. Memory arrays (e.g. RAM, ROM, PLA, etc.) are more susceptible to noise margin reductions. To make them more robust, they should be designed to be either differential, or full swing. In the prototype system (Chapter 6), the only types of memory arrays used were RAMs and CAMs, which were designed with differential bitlines for improved robustness. The critical circuits which are most susceptible to noise margin reduction are the I/O transceivers due to their very large, and localized, peak currents.

Timing Violations

As described in Section 4.4.4, local on-chip supply variations can lead to timing violations if a critical path sufficiently slows down to exceed the clock cycle time of the ring oscillator. A DVS processor cannot have timing violations induced by global supply variations, since the ring oscillator's delay (e.g. the inverse of the clock frequency) will vary with the delay of the critical paths.

A design margin of 5% was included in the timing verification of the processor to account for localized voltage drops. The equation for delay sensitivity (Equation 5.12), which is the relative change in delay (V_{DD}) for a given ΔV_{DD}, can be rewritten to translate this 5% design margin into a maximum allowable voltage drop, ΔV_{DD}, at a given value of V_{DD}:

Power Distribution Methodology

$$\Delta V_{DD} \approx \frac{\frac{\partial \text{Delay}}{\text{Delay}} \cdot \text{Delay}}{\partial \text{Delay} / \partial V_{DD}} = \frac{5\% \cdot \text{Delay}}{\partial \text{Delay} / \partial V_{DD}} \quad \text{(EQ 5.12)}$$

and this can be used to calculate the maximum resistance, R_{MAX}, allowed on the supply line given a gate's peak output current, I_{GATE}, which is also a function of V_{DD}.

$$R_{MAX} \approx \frac{5\% \cdot \text{Delay}(V_{DD})}{\partial \text{Delay} / \partial V_{DD}} \cdot \frac{1}{I_{GATE}} \quad \text{(EQ 5.13)}$$

If the supply distribution network is designed so that the resistance that the gate sees to a solid, reference voltage is less than R_{MAX}, then any delay variation due to supply variation will fall within the 5% design margin, and the processor will continue to operate correctly without any spurious timing violations.

5.4.2 Chip-level Distribution

The chip-level power distribution network for the processor chip is shown in Figure 5.14. The other DVS-compatible chips in the processor system (SRAM and I/O chips) were designed in similar fashion, but the processor chip is focused on in detail to demonstrate the power distribution methodology. There are two separate power supplies, V_{DD} for the core, and V_{DDIO} for the pad ring of the chip, to isolate the core circuitry from the I/O transceivers. There is a single ground on the chip, since the low-impedance substrate of our process makes separate ground supplies difficult to isolate.

There were two primary design goals for the power distribution network. First, there should be sufficient bypass capacitance on the chip to supply the charge for the large switching currents. This will reduce the voltage drop across the bonding wires and packaging pins, which would otherwise be intolerable. This is not an issue for traditional low-power/low-voltage chips, but it is an issue for a DVS-compatible chip which can have large switching currents at high voltage. The second goal was to ensure that any point on the chip has a low-impedance connection to a solid voltage reference, either a large on-chip bypass capacitor or the external voltage supply. This was critical to eliminate timing violations due to localized voltage drops.

In the 3 level metal technology used, ground is routed in *Metal3* directly over power in *Metal2*, except where the power lines dissect a block, in which they are routed in

Metal3 and *Metal1*, leaving *Metal2* to be used for signal routes. Routing ground on top of power helps to both maximize the inter-metal bypass capacitance, and minimize inductive losses via magnetic field cancellation.

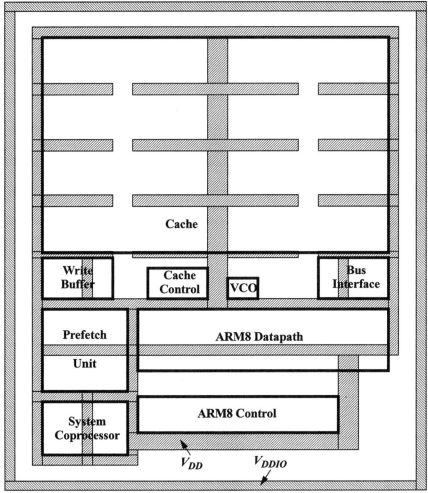

FIGURE 5.14 Prototype Processor IC Power Distribution Network.

A total of 16 pins for V_{DD} and another 10 pins for V_{DDIO} are distributed uniformly around the periphery of the chip. The large number is essential to reduce the inductive

Power Distribution Methodology

voltage drop on the bonding wires. Additionally, by spreading them evenly around the chip's edge, any point on the V_{DD} and V_{DDIO} networks sees minimal resistance to the off-chip power supply.

Bypass Capacitance

The bulk of the bypass capacitance is provided by NMOS devices which provide the highest capacitance per area. These devices were placed in all the significantly-sized open areas of the chip as well as under the power lines themselves. Power is connected to the gate, while the source and drain are tied to ground, maintaining the device in the triode region of operation.

The width and length of the NMOS devices are constrained in size due to resistance in the device channel and in the polysilicon gate. The maximum RC time-constant of the channel is to its mid-point, and is:

$$\tau_{CHAN} = \frac{1}{4} \cdot R_{DS} \cdot C_{GS} = \frac{1}{4} \cdot \frac{(W \cdot L \cdot C_{OX})}{k_p \cdot (W/L) \cdot (V_{GS} - V_T)} = \frac{L^2 \cdot C_{OX}}{4 \cdot k_p \cdot (V_{DD} - V_T)}$$
(EQ 5.14)

where the factor of four occurs because the source and drain are connected, dividing the effective R and C by a factor of two each. The time-constant scales inversely with V_{DD} similar to circuit delay, so that the maximum channel length is not strictly limited at maximum supply voltage and current draw, but relatively optimal over all V_{DD}. The maximum time-constant of the polysilicon gate is:

$$\tau_{GATE} = (R_{GATE} \cdot C_{GS})/4 = \frac{1}{4} \cdot \left(\rho_{PLY} \cdot \frac{W}{L}\right) \cdot (W \cdot L \cdot C_{OX}) = \frac{\rho_{PLY} \cdot W^2 \cdot C_{OX}}{4}$$
(EQ 5.15)

where ρ_{PLY} is the sheet resistance of polysilicon (10 Ω/sq. in our 0.6μm process) and the factor of four occurs because the gate is contacted on both sides of the device. Unlike τ_{CHAN}, τ_{GATE} is independent of V_{DD} so the maximum channel width is strictly limited at maximum voltage.

For the core circuitry (powered by V_{DD}), the current spikes are on the order of 50-100ps wide at maximum voltage. To make the RC delay of the bypass capacitance negligible, the maximum time-constant value was set to 25ps. This dictated the maximum W/L of a bypass NMOS device to be 54/3 μm in our 0.6μm process. When

Energy Driven Design Flow

including the overhead for routing the bypass capacitors, the area efficiency of this size bypass device is approximately 60%.

Since all the power metal lines shown in Figure 5.14 contain bypass capacitors underneath them, and the network is connected with cross-bar metal lines between blocks, the entire V_{DD} network provides a solid voltage reference. That is, the voltage on the V_{DD} network varies approximately uniformly across the chip.

Providing a solid voltage reference for V_{DDIO} is a much more difficult task. For the worst-case condition that all I/O pins transition low to high, and they drive the maximum 50pF external load capacitance, the total capacitance charged in a cycle can be as high as 2nF. Bypass capacitance is used to mitigate the voltage drop on V_{DDIO}, placed both under the V_{DDIO} power lines and in all available open areas between pads. But, only 2nF of capacitance could be placed on-chip, such that for the worst-case condition, the on-chip bypass capacitance cannot supply all of the charge, leading to localized voltage drops on the I/O transceivers.

However, the bus interface was designed to have the minimum amount of gates in its path delay to provide the requisite design margin for very large increases in delay driving the external bus. By eliminating the bus interface as a critical path, large voltage drops on V_{DDIO} are tolerable without inducing timing violations. Thus, only the core circuitry connecting to the V_{DD} network had to be carefully designed to guarantee no timing violations by design.

Local Supply Routing

As was demonstrated in Section 4.4.4, the delay sensitivity to V_{DD} is a maximum at $2 \cdot V_T$ (approximately 2V in our 0.6μm process), so the local supply network must be designed at this value of V_{DD}, which will determine the smallest amount of resistance tolerable in the power supply network. At 2V, Equation 5.13 is roughly $0.05/I_{GATE}$. This equation can be expressed in terms of NMOS device width (W_N) and number of gates in parallel (N_{GATE}) for 0.6μm process as:

$$R_{MAX} \approx \frac{0.05}{N_{GATE} \cdot v_{SAT} \cdot C_{OX} \cdot W \cdot (V_{DD} - V_T - V_{Dsat})} = \frac{0.001}{N_{GATE} \cdot W_N} \quad \text{(EQ 5.16)}$$

The equivalent resistance for a PMOS device is for a device with a width is 2.5 times as large as the NMOS device due to the decreased mobility.

This equation was used to determine how wide the local power routes to the core circuitry should be. For the 32-bit datapaths of the ARM8 core, the prefetch unit, and the coprocessor, the 100µm-wide *Metal2* and *Metal3* lines of the power distribution network are spaced 825µm apart as shown in Figure 5.15. Within the datapath, *Metal2* and *Metal3* are used for signal routing such that power had to be routed to the individual bits in parallel *Metal1* lines. The maximum resistance to the power distribution network is from the midpoint of the datapath, whose resistive losses can be modeled as shown.

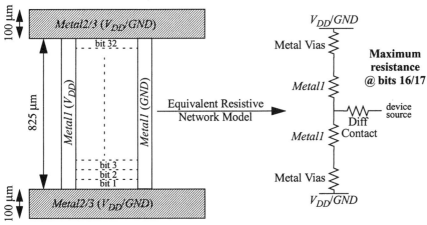

FIGURE 5.15 Local Datapath Power Routing.

For a simple datapath cell with minimal output load (a 1x gate), $W_N = 1.2$µm. A worst-case analysis must assume all 32 bits of the datapath switch at once, which yields $R_{MAX} = 26\Omega$. The contacts, vias, and metal wires must be designed so that:

$$R_{MAX} > R_{DIFF} + \frac{1}{2}(R_{MET1} + R_{VIAS}) \qquad \text{(EQ 5.17)}$$

For our process technology, R_{DIFF} will always be approximately 20% of R_{MAX}, because the number of parallel diffusion contacts scales with W_N, so that R_{DIFF} scales down with R_{MAX}. The metal via resistance, R_{VIAS}, can be made negligible (<2% of R_{MAX}) because many vias can be added in parallel under the 100µm wide power network. Thus, the width of Metal1 must be adjusted so that:

Energy Driven Design Flow

Energy Driven Design Flow

$$R_{MET1} < 1.6 \cdot R_{MAX} = \frac{50\mu}{W_N} \qquad R_{MET1} = \frac{L_{MET1}}{W_{MET1}} \cdot \rho_{MET1} = \frac{31\mu}{W_{MET1}} \qquad \text{(EQ 5.18)}$$

and combining these two relations for R_{MET1} gives:

$$W_{MET1} > 0.6 \cdot W_N \qquad \text{(EQ 5.19)}$$

which is a concise rule-of-thumb for sizing the *Metal1* power lines given the size of a gate's transistors. By following this, the datapath could be implemented to not have any potential timing violations, due to power line resistance.

Similar equations were developed for the large cache memory arrays, as well as the placed-and-routed standard cells of the core control logic. For the standard cell arrays, the cell-level power was routed in *Metal1*, as shown in Figure 5.16, again to free up

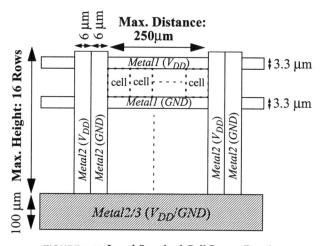

FIGURE 5.16 Local Standard Cell Power Routing.

Metal2 and *Metal3* for signal routing. However, to reduce resistance to the power distribution network, as necessary, 6μm-wide *Metal2* straps connected up the power lines, and could be spaced no further than 250μm apart. The maximum number of rows was calculated to be sixteen. Again, by following these constraints, any place-and-routed standard cell implementation could be guaranteed to not have potential timing violations by design.

5.4.3 Board-level Distribution

The DVS regulator chip[5.9] is very sensitive to board-level parasitics which may interact with the output LC filter for the DVS supply voltage, V_{DD}. Thus, careful attention had to be paid in laying out the power distribution for the board. An entire PCB plane was dedicated to V_{DD}, which reduced the parasitic resistance to a negligible amount. This also allowed a single PCB via to connect the V_{DD} pins on the chip sockets to the power plane, in order to minimize the inductance.

Since the plane provides negligible resistance and inductance, the need to place the filter capacitor next to the inductor was eliminated. Instead, the 5µF capacitor was evenly spread out across the board, and placed next to the chips' V_{DD} pins as 100nF and 200nF bypass capacitors. This provided both the necessary filter capacitance, as well as good bypass capacitance, to eliminate unnecessary noise on the chips' V_{DD} pins.

5.5 Functional Verification

At every stage in the design process, the design specification was verified for functional correctness. At the higher, more abstract, design levels, logical behavior is checked. Closer to the physical design, individual signals are checked for phase correctness and setup/hold times in additional to logical behavior. A verification methodology was developed so that test code had to be developed only at the top design level. At all subsequent design levels, scripts automatically generated new test vectors from the previous design level's test vectors.

5.5.1 Behavioral Verification

The essence of behavioral verification is ensuring that the processor system properly operates from the programmer's point-of-view. The master reference for comparing against is the specified behavior of the Instruction Set Architecture (ISA). For example, an add instruction that adds two register and places the result in another register must be validated for all specified register combinations as well as all register values. Since exercising all possible combinations would create unrealizable test code, the approach is to heuristically deriving tractable test code that covers all the "significant" state transitions.

In addition, the processor requires a coherent memory hierarchy. The physical memory hierarchy may store the same address contents in multiple locations, such as in the cache and in main memory, but these contents must be consistent. The memory hierarchy must also provide a minimum amount of memory management, such as trapping illegal memory accesses, to prevent lock of the processor. Hence, the physical memory hierarchy must be verified to be logically correct over a wide range of possible operations.

ARM Ltd. provided both the ARM8 processor core behavioral model and its suite of validation test code *[5.2]*. By restricting any changes to the behavior model, the validation suite could be used verbatim for verifying the processor core and its implementation of the ARM v4 ISA. Test code had to be developed for the rest of the microprocessor and external system components.

The basic design flow to create this test code is shown in Figure 5.17. This verification flow ran in conjunction with the high-level behavioral design once enough of the system was specified to be simulatable. Individual code was developed to test each major part of the microprocessor (e.g. System Coprocessor, Cache Controller, Write Buffer, etc.) as well as the external SRAM chip and I/O chip. To guarantee correctness by design, the code was written to be self-checking; the result of any operation under test would be compared against the expected value, and flag an error on a mismatch. Both the cycle-level and VHDL simulator models included basic I/O so that error messages could be printed to the simulator screen.

The self-checking test code enforces consistency between the C simulator and the ISA, as well as consistency between the C simulator and VHDL behavioral model. The golden test code suite consists of: 28 programs from ARM Ltd. that test individual ARM8 blocks, 32 programs from ARM Ltd. which comprise their validation suite to verify the entire ARM8 core, and an additional 10 programs that were developed for the remainder of the system.

Once all the test code successfully ran on the VHDL system simulation, there was extremely high confidence in the functional correctness of the behavioral model. The additional step taken was to boot the operating system on the VHDL simulation to further verify functionality.

5.5.2 Test Vector Generation

Once the golden test code suite was developed, test vectors for individual blocks could be generated from a behavioral VHDL simulation as demonstrated in

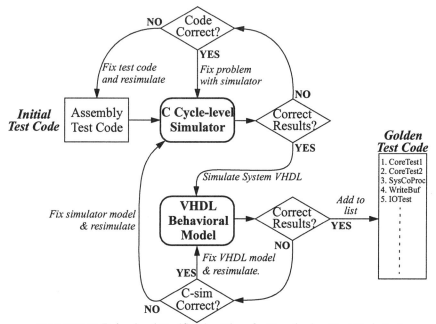

FIGURE 5.17 Behavioral Verification Flow for Developing Test Code Suite.

Figure 5.18. The block under test (BUT) has its pins traced while running the test code that exercises the BUT. The output waveform database is then sampled and converted to time-independent test vectors via the *leap2crf* script. Each vector corresponds to a single clock phase, gives input state at the beginning of the phase, and indicates what the output state should be at the end of the phase.

A header file describes the direction of each pin to be input, output, or bidirectional. In the test vector file, a bidirectional signal is treated either as an input or output for any given test vector, depending on whether the signal was actively driven by the BUT in that test vector or not. An *H* or *L* state indicates a driven input, while a *1*, *0*, or *Z* state indicates an output signal. A skew file provides a skew number for each signal, which can be used to model setup and hold time constraints on the BUT.

The golden test vectors are used directly as structural VHDL test vectors. A script was written which allows switch-level simulation test vectors to be automatically generated from the golden test vectors as well as the expected response. This script takes a time base as an argument so that simulation test vectors can be created at arbitrary cycle times allowing vector to be quickly generated for multiple voltages. When

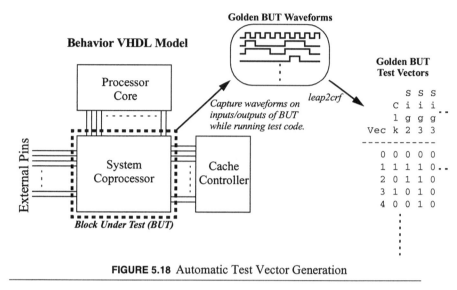

FIGURE 5.18 Automatic Test Vector Generation

setup and hold time constraints had to be renegotiated between blocks, the blocks' skew files were modified and adjusted switched-level simulation test vectors could be automatically generated.

A hierarchal set of test vectors were generated. Some vectors were for smaller blocks, such as an individual ARM8 macro block or cache controller, and others were for larger blocks, such as the entire ARM8 core, the cache subsystem, and even the entire microprocessor chip. This allowed the design to be verified at all levels of the hierarchy. Most of the logical debugging occurred within the smaller blocks, while verifying the larger blocks ensured the design was connected properly.

5.5.3 Structural Simulation

A script was used to create a testbench for a given block. The block's VHDL entity provides the necessary pin-direction information, and creates a VHDL stimulus file which reads the test vector file. At the beginning of each vector, the inputs are toggled as specified, and at the end of the specified half-cycle time, the outputs are checked against the expected outputs in the test vector file. If any errors are present, an error flag is issued giving the exact location of the error.

As the design was refined from a high-level behavioral specification to a lower-level structural VHDL specification, the test-bench simulation was run for each block to ensure proper functionality. If all test vectors passed successfully, the block was deemed to be functionally correct and ready for physical implementation.

5.5.4 Transistor Netlist Simulation

TimeMill, a commercial switch level timing simulator, was used for transistor level simulation. An input vector file and an output vector check file were generated from the behavioral description as described in Section 5.5.1. The input vector file specifies input signal transitions at regular time intervals, equal to the desired half-cycle time. The skew file can be used to shift signal transitions with respect to the clock edge. The output vector check file specifies what the output pin states should be at regular intervals. Again, with the skew file, when the outputs should be stable with respect to the clock edge can be varied with setup-time constraints.

The initial simulation is run on a block's schematic netlist, to ensure topological functionality. This step catches logical transistor design errors. Once the block's layout is complete and verified against the schematic via LVS, the layout was extracted to create a netlist. TimeMill was run on this new netlist, and would catch setup and hold time violations for the block due to the interconnect parasitics.

5.6 Timing Verification

There are two parts to timing verification: race-condition analysis for functionality, and critical-path analysis for performance. PathMill, a commercial static timing analyzer, was extensively used for timing verification, in which the key to speeding up the analysis turn-around was a schematic design that followed a simple naming methodology. Then, PathMill could be scripted to quickly find the important paths to be analyzed.

Timing was first run on the schematic design to flag potential race paths as well as identify critical paths based upon logic depth. Any path longer than 30 gate delays was reduced. Paths in the range 24-30 gates deep were reduced if a simple schematic fix was possible, since they had a high probability of becoming critical with extracted parasitics. Any path depth less than four was increased, which would occur as the unintended side-effect of a sped-up latch. Finally, any path that could induce a clock glitch (e.g. an output signal from a Phi2 latch gating a Phi2 clock driver) was flagged and fixed.

Once the first-pass layout was complete, PathMill was run again on the extracted netlist. Multiple timing iterations were used to fix critical paths that cropped up due to lack of interconnect capacitance in the schematic netlist. A script was written and used to ensure that the layout netlist had the exact node names as the schematic netlist. Hence, the exact same timing input deck could be used both on the schematic and extracted netlists.

5.6.1 Schematic Naming Methodology

All cell instances that either maintain state (e.g. latches, flip-flops, registers) or are clock drivers have specific naming requirements. The name conveys information on the type of state and on which phase, 1 or 2, the instance is active. Within the master instance, the actual state nodes also require specific names. Then, for example, through simple pattern matching all phase-2 latch nodes can be specified in one statement.

The special instance labels are:

$p\{1,2\}lat\{\#\}$	Phase-1/2 transparent latch.
$p\{1,2\}flop\{\#\}$	Phase-1/2 edge-triggered flip-flop.
$p\{1,2\}clk\{\#\}$	Phase-1/2 clock driver.
$p\{1,2\}enab\{\#\}$	Phase-1/2 enabled signal (tri-stated).
$p\{1,2\}pre\{\#\}$	Phase-1/2 pre-charged node.

The $\{\#\}$ is used to differentiate multiple instances of the same type/phase on a given schematic sheet (e.g. *p1lat0*, *p1lat1*, etc.). In addition, there is the special phase designator, *A*, indicating an asynchronous signal (e.g. *pAenab0*).

The special state node labels are:

epic_latch	Latch node between pass gate and cross coupled inverters.
epic_latch{1,2}	The two latch nodes in a flip-flop.
epic_clock	Clock node right after the enabled gate cell in the clock drivers.
epic_enable	Node immediately preceding the driving inverter on a tri-state bus. Inverter enabling signal must be the output of a clock driver.

epic_pre   Precharged node. Precharge signal must be the output of a clock driver.

epic_register   Latch node within a register file. Labelled differently because it requires special handling since the register files are bi-phase.

In addition, any one of the above state nodes can be suffixed with a letter indicating parallel nodes on a single schematic sheet (e.g. *epic_latcha*, *epic_latchb*).

Through wild-card matching, all phase-2 latch nodes are simply **p2lat*.epic_latch**. To ensure all instances are properly labelled, a script was written to parse through the netlist looking for any node names not associated with a labelled instance, and flag them to be changed. Likewise, the inverse can be checked as well. However, if both the instance and node labels are omitted, then the node cannot be caught just from parsing the netlist. However, it will most likely show up as a critical path, and can be changed after the initial timing run. Thus, this methodology proved very robust in properly annotating all state nodes.

5.6.2 Path Identification

When performing timing analysis, all delay measurements must be referred back to a common signal, typically the output of the global clock driver. As was demonstrated in Section 5.3.4, the maximum skew between any two clock driver outputs is less than 4% of the cycle time. Because all the state nodes (except *epic_register*) only change state upon a clock driver output transition, timing between state nodes can ignore delay through the clock drivers, with a 4% margin of error.

In Figure 5.19, when *Clock* goes low, *Phi1/nPhi1* get asserted, latching a new data value onto the state node, *epic_latcha*. When *Clock* goes high, *Phi2/nPhi2* get asserted, latching a new value onto the state node *epic_latchb*, at which time, the input to *p2lat* must be stable. To ensure this, the delay through the logic block, t_{CRIT}, referred back to the common *Clock* signal must be less than the target cycle time:

$$t_{CRIT} = t_{d1} - t_{d2} = t_{Clock \rightarrow epic\_latcha} + t_{epic\_latcha \rightarrow epic\_latchb} - t_{Clock \rightarrow epic\_latchb} \quad \text{(EQ 5.20)}$$

where the delay through the clock buffers is:

$$t_{Clock \rightarrow epic\_latch\{a,b\}} = t_{Clock \rightarrow Phi\{1,2\}} + t_{Phi\{1,2\} \rightarrow epic\_latch\{a,b\}} \quad \text{(EQ 5.21)}$$

Energy Driven Design Flow

Energy Driven Design Flow

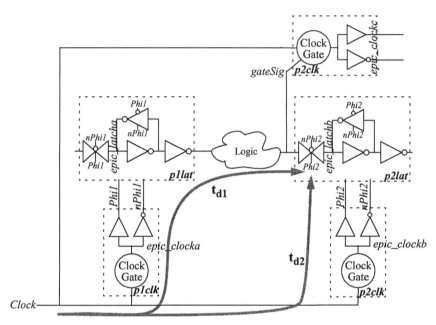

FIGURE 5.19 Simple Circuit Example.

The first component differs at most by the maximum clock skew. The second component will be essentially the same between the two because the output of the clock drivers have nearly identical rise/fall times. Hence, Equation 5.20 can be rewritten as:

$$t_{CRIT} = t_{\text{epic\_latcha} \to \text{epic\_latchb}} + \delta_{\text{ClkSkew}} \cong t_{\text{epic\_latcha} \to \text{epic\_latchb}} \quad \textbf{(EQ 5.22)}$$

Hence, timing path analysis only requires looking between any two state nodes (latch, flip-flop, precharged, and enabled), which is virtually equivalent to referring the path calculation back to a common point. There are two exceptions to this rule.

The first exception is for clock gate signals. For the path from *p1lat* to *p2clk*:

$$t_{CRIT2} = t_{\text{Clock} \to \text{epic\_latcha}} + t_{\text{epic\_latcha} \to \text{gateSig}} + t_{\text{gateSig} \to \text{epic\_clock}} - t_{\text{Clock} \to \text{epic\_clock}} \quad \textbf{(EQ 5.23)}$$

The last term is dropped, because the *epic_clock* node should be stable before *Clock* changes. Equation 5.23 can be reduced to:

$$t_{CRIT2} = t_{\text{Clock} \to \text{epic\_latcha}} + t_{\text{epic\_latcha} \to \text{epic\_clock}} \quad \text{(EQ 5.24)}$$

Thus, the delay between the state node (*epic_latcha*) and the clock node (*epic_clock*) must be less than the target cycle time minus the delay through the clock buffer. The mean delay through the clock buffer is 8% of the half-cycle time of over voltage. Thus, these paths have a shorter time to complete.

The other exception are *epic_register* nodes. All the register files are bi-phase, so the standard clock drivers cannot be used. Instead, custom clocking circuitry is required, which while designed to match the clock driver as well as possible, they do not meet the same skew tolerance. Thus, input and output to these nodes must be given an extra margin of tolerance, and carefully simulated to ensure no race condition exists.

5.6.3 Timing Analysis

Path delay varies from VCO delay in one of two ways. First, there are paths whose delay is dominated by PMOS devices (since $V_{TP} > V_{TN}$) and/or interconnect capacitance, and they slow down at low voltage with respect to the VCO. The second class of paths, whose delay is dominated by gate/diffusion capacitance (which increases with voltage) and/or interconnect RC delay, slow down at high voltage with respect to the VCO. A typical commercial design is only concerned about a singular voltage operating point, but DVS must operate over a broader range of voltage.

However, DVS only requires timing analysis at two voltages, which are the extremes of the desired operating range. For the prototype processor, the voltages are 3.3V and 1.2V. If timing constraints are met at these two points, then the timing constraints will be met at all intermediate points in between.

Using the schematic naming methodology, a complete timing analysis can be performed with 16 individual PathMill runs, listed in Table 5.6 . This analysis checks for short paths, long paths, and illegal paths. The target cycle time is 10ns at 3.3V, and 80ns at 1.2V.

Checks 1-4 analyze paths between different-phase state nodes, so they only have a half-cycle time to complete. In addition, the maximum delay is further reduce by 5% of the cycle-time to provide margin for clock skew. The first two checks are soft constraints, because the sink state node is transparent, and will continue to operate if the maximum delay is exceeded. Checks 3-4 are hard constraints, because a functional error will result if the maximum delay is exceeded. Checks 5-6 have a full cycle to complete minus 5% margin for clock skew. Checks 7-8 provide 10% margin to account for both clock skew and delay through the clock buffer. The register checks (9-12) analyze both input/output paths, and provide 10% margin to give additional headroom for the custom clock drivers. The last four checks (13-16) check for illegal paths that should never occur.

TABLE 5.6 Timing Analysis Pattern Set

#	Source	Sink	Max Delay(ns) V_{DD}=3.3	Max Delay(ns) V_{DD}=1.2
1	p1{lat,flop,enab,pre}	p2{lat,enab}	4.5 (soft)	36 (soft)
2	p2{lat,flop,enab,pre}	p1{lat,enab}		
3	p1{lat,flop,enab,pre}	p2{flop,pre}	4.5 (hard)	36 (hard)
4	p2{lat,flop,enab,pre}	p1{flop,pre}		
5	p1{lat,flop,enab,pre}	p1flop.epic_latch1	9.5 (hard)	76 (hard)
6	p2{lat,flop,enab,pre}	p2flop.epic_latch1		
7	p1{lat,flop}	p2clk.epic_clock	4.0 (hard)	32 (hard)
8	p2{lat,flop}	p1clk.epic_clock		
9	p1{lat,flop,enab,pre}	epic_register	4.0 (hard)	32 (hard)
10	p2{lat,flop,enab,pre}			
11	epic_register	p1{lat,flop,enab,pre}	4.0 (hard)	32 (hard)
12		p2{lat,flop,enab,pre}		
13	p1{lat,flop,enab,pre}	p1clk.epic_clock	Should never occur	
14	p2{lat,flop,enab,pre}	p2clk.epic_clock	(same phase gate signal)	
15	p1{lat,flop,enab}	p1pre.epic_pre	Should never occur	
16	p2{lat,flop,enab}	p2pre.epic_pre	(same phase pre input)	

The output from PathMill provides an ordered list of long and short paths for each check. In addition, PathMill reports each stage in these paths, such that they can be quickly identified in the schematic. If any paths exceed the maximum delay check, they are fixed through schematic changes, and the design is re-analyzed.

References

[5.1] Advanced RISC Machines, Ltd., *ARM Architecture and Implementation Reference*, Document Number ARM-DDI-0100A-I, Feb. 1996.

[5.2] Advanced RISC Machines, Ltd., *ARM Software Development Toolkit Reference Guide*, Document Number ARM-DUI-0041A, Jan. 1997.

[5.3] P. Landman, J. Rabaey, "Black-Box Capacitance Models for Architectural Power Analysis", *Proceedings of the 1994 International Workshop on Low-Power Design*, Napa Valley, CA, April 1994.

[5.4] T. Burd, *Low-Power CMOS Library Design Methodology*, M.S. Thesis, University of California, Berkeley, Document No. UCB/ERL M94/89, 1994.

[5.5] N. Weste and K. Eshraghian, *Principles of CMOS VLSI Design*, Addison Wesley, Reading, MA, 1993.

[5.6] Advanced RISC Machines, Ltd., *ARM Architecture and Implementation Reference*, Document Number ARM-DDI-0100A-I, Feb. 1996.

[5.7] D. Markovic, *Analysis and Design of Low-Energy Clocked Storage Elements*, M.S. Thesis, University of California, Berkeley, Document No. UCB/ERL M00/64, 2000.

[5.8] J. Yuan, C. Svensson, "High-Speed CMOS Circuit Techniques", *IEEE Journal of Solid-State Circuits*, Vol. 24, No. 1, Feb. 1989

[5.9] A. Stratakos, *High-Efficiency, Low-Voltage DC-DC Conversion for Portable Applications*, Ph.D. Thesis, University of California, Berkeley, 1998.

CHAPTER 6 Microprocessor and Memory IC's

6.1 Microprocessor IC

The chip's processor core implements the ARM V4 instruction set architecture (ISA) *[6.1]*. The implementation was derived from an RTL behavioral model (provided by ARM Ltd.) which fixed both the ISA as well as the processor core interface. However, both the custom physical implementation of the core, as well as the rest of the microprocessor design, were fully optimized for energy efficiency.

Full compatibility of the ISA was critical so that commercially-available compilers, assemblers, and simulators could be used, thereby allowing rapid software development on the hardware platform, and eliminating the need to develop custom software tools. While the microprocessor implementation is ARM-based, the design methodology outlined in the previous chapters is equally applicable to other ISAs, and will yield similar improvement in energy efficiency.

Inter-chip communication is much more costly than intra-chip communication, in terms of both performance and energy efficiency, so that integrating as much system functionality as possible on the microprocessor chip will yield a more energy-efficient implementation. As such, all system logic was integrated onto the microprocessor, with the exception of the main memory, the voltage regulation loop, and the I/O interface. The main memory remained separate because sufficient amounts of memory cannot be integrated onto the same die. The loop and interface remained separate for design simplicity, without having a significant impact on energy efficiency. The microprocessor die, shown in Figure 6.1, measures 7.5 x 9.0mm and contains 1.3M transistors of which 890k are memory (CAM & SRAM) transistors.

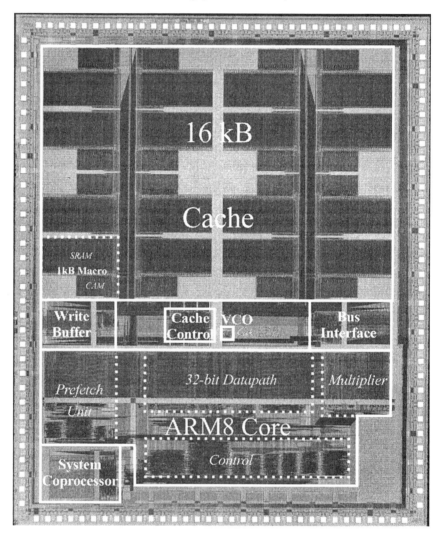

FIGURE 6.1 DVS Microprocessor .IC

6.2 Processor Architecture

A high-level block diagram of the microprocessor's architecture is shown in Figure 6.2. The processor core is a custom implementation of an ARM8, which is a 32-bit five-stage scalar integer pipeline with an eight-word prefetch unit that performs

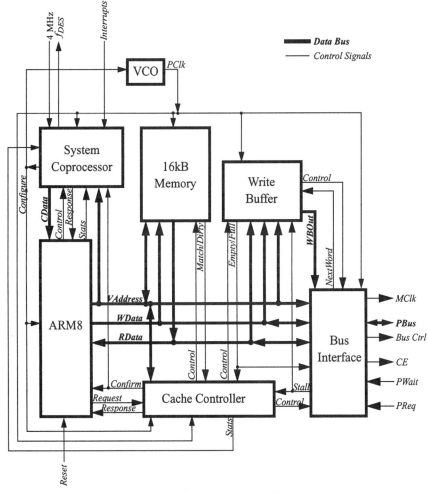

FIGURE 6.2 Microprocessor Architecture.

static branch prediction *[6.2]*. The system coprocessor contains system control state which can be manipulated by the ARM8 core via its coprocessor interface. The cache sub-system consists of a 32-way set-associative 16kB memory, a twelve-element write buffer, and a bus interface, all of which are managed by the cache controller. The cache is physically indexed to eliminate the need for a TLB, and the upper six bits of the address are utilized for memory-space control which gives the microprocessor an effective 26-bit address space. The bus interface drives the external system bus, and contains a memory controller which allows external memory chips to be seamlessly connected to the bus. The VCO provides the variable-frequency clock signal, *PClk*, to all the internal microprocessor components, and *PClk* is buffered and transmitted to the converter as the f_{CLK} signal. The external bus is clocked by *MClk*, which is divided down from *PClk*.

Data Flow

The data flow of the microprocessor was designed around the fixed ARM8 interface. The memory interface consists of three unidirectional busses; the address bus (*VAddress*), the write data bus (*WData*), and the read data bus (*RData*).

The system coprocessor sends data to the ARM8 on a dedicated unidirectional bus (*CData*). The coprocessor receives data via *VAddress*, and during this transfer cycle, the ARM8 must suppress any pending memory access and place the data word on *VAddress*. Since coprocessor writes are infrequent, this has negligible impact on processor performance due to the forced cancellation of memory-access cycles.

VAddress is an input to the write buffer and the bus interface, and is an input/output to the cache memory so that the tag array can be read and written. The cache controller also reads and writes the lower bits of *VAddress* so that it can detect cache-line boundaries and generate cache memory block enables on reads, and increment *VAddress* across a line for cache-line loads and write-backs.

WData is an input to the cache memory and the write buffer. The bus interface has a bidirectional *WData* port so that it can input non-cacheable ARM8 writes to be sent to external memory, and output data to the cache memory during cache-line loads. *RData* is an output of the cache memory and an input to the write buffer. The bus interface *RData* port is also bidirectional so that it can input data from the cache memory during cache-line write-backs to external memory, and send data to the core for non-cacheable ARM8 reads.

Since the system data bus is a single, bi-directional bus (*PBus*) which carries time-multiplexed address and data words on it, as described further in Section 9.1, the bus

Processor Architecture

interface multiplexes the three internal busses onto *PBus*. Due to this multiplexing, the write buffer stores both address and data words into a single twelve-element queue, whose contents are sent to the bus interface via a dedicated unidirectional bus (*WBOut*).

Clock Control Domains

To eliminate unnecessary clocking and circuit activity, there are three top-level clock control domains as shown in Figure 6.3. The core domain contains the ARM8 core and system coprocessor, which fetch and execute instructions when this domain is active. The ARM8 expects read and write memory accesses to complete in a single cycle, and communicates these accesses to the cache controller via the ARM8's memory request and response handshake protocol *[6.2]*. When the cache sub-system cannot complete the request in one cycle (e.g. cache miss, full write buffer, blocked bus interface, etc.), the cache controller halts the core domain via the *Confirm* signal, which gates *PClk* within all of this domain's clock drivers.

The cache domain contains the entire cache subsystem with the exception of the system-bus side of the bus interface. The cache controller directs the operation of the cache memory, write buffer, and bus interface via various control signals. In response

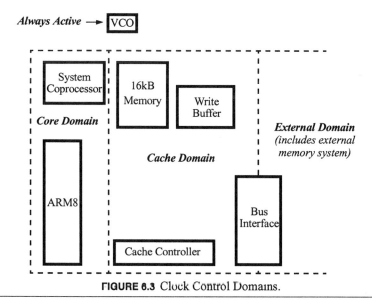

FIGURE 6.3 Clock Control Domains.

Microprocessor and Memory IC's

to a memory request from the core, the cache memory's tag array returns whether the request is a match, and if it is not, the cache controller initiates a cache-line load. If the access is not a match and the cache line is dirty (i.e. it has been written to in the cache, but not updated in main memory), a cache-line write-back must be executed first before initiating the cache-line load. The cache controller routes all bufferable writes to the write buffer. When the buffer is full during a pending write, then the cache controller halts the processor and waits for a slot to open. Similar to the ARM8 memory interface, the cache controller expects a read or write access to the bus interface to complete in a single cycle. Since the system bus can operate at no more than one-half the internal clock speed, the bus interface will halt the cache controller, via the *Stall* signal, until the access is complete. This in turn halts the core domain's clock, as well.

The external domain is clocked by *MClk*, and encompasses the system-bus side of the bus interface, as well as the external memory system. An asynchronous interface connects the core-side of the bus interface, which is manipulated by the cache controller and clocked by *PClk*, to the bus-side of the bus interface, which connects directly to the external system bus and generates the external memory chip enables (*CE*). For external memory reads and I/O accesses that require multiple *MClk* cycles, external chips can stall the bus-side of the bus interface via *PWait*, which also halts the first two clock domains as well since they are both waiting for the pending system bus access to complete.

Write Buffer Control Flow

When there is no pending external-memory access request from the cache controller in a given cycle, the bus interface polls the write buffer to see if is not empty. When it is non-empty, the bus interface will autonomously initiate a system bus access to write out the data to the external memory system. During this transfer, if the cache controller has an access request to the bus interface (e.g. non-cacheable memory request from the ARM8, cacheable request that takes a cache miss, etc.), the cache controller must wait for the transfer to complete and stall the core clock domain. To ensure memory consistency without additional hardware, any ARM8 read request must stall until the write buffer is empty *[6.3]*. With the large 16kB cache, this condition is infrequent and stalling the core has negligible impact on processor performance.

DMA Control Flow

The I/O chip can directly access the main memory via a direct memory access (DMA). A DMA request is sent to the microprocessor via the *PReq* signal. When there is no outstanding system bus transfer, the bus interface grants the DMA request and releases control of the system bus to the I/O chip. Until the DMA request completes, any access request to the bus interface from the cache controller is stalled.

Processor Configuration & Monitoring

The system coprocessor, described further in Section 6.2.5, is responsible for configuring the microprocessor and collecting dynamic statistics. The coprocessor sets the processor speed by transmitting f_{DES} to the converter chip, and controls the voltage-to-frequency conversion by configuring the VCO. In addition, it can configure the ARM8 core, as well as the operation of the cache subsystem via the cache controller. The coprocessor collects run-time statistics from both the ARM8 and cache controller, which can be accessed in software via a coprocessor read instruction.

6.2.1 Processor Core

The processor core is a fully-compatible, custom-implementation of the commercial ARM8 core. Starting from a block level RTL behavioral description, the design methodology described in previous chapters was utilized to provide an energy-efficient and DVS-compatible implementation. This section provides an overview of the ARM8 core and highlights some of the specific design optimizations. A more detailed description of the core's functionality, I/O interface, and signal timing can be found in the ARM8 data-sheet from ARM Ltd. *[6.2]*

ARM8 Instruction Set Architecture

The ARM8 is similar to a traditional RISC ISA as it is a load-store architecture. Data processing instructions can only operate on registers; external memory contents can be loaded to and stored from the register bank, but not operated on directly. In addition, the instructions are a fixed size of 32 bits. These characteristics allow the ISA to map onto the common five-stage pipeline found in simple RISC processor cores.

There are some non-traditional features of the ARM8 ISA which prove useful in embedded applications by reducing the machine code size. However, these add complexity to the pipeline control and data flow. The primary features are:

> 1) All instructions are conditionally executed. Each instruction has a four-bit condition field, which must be evaluated before writing the results of the instruction to the register bank or main memory, or passing the results to subsequent instructions via data forwarding.

2) The second operand of data processing instructions can be shifted before the data processing operation. A five-bit field specifies the shift amount, and the shift type can be logical left, logical right, arithmetic right, and rotate right. Because of this feature, there is no explicit shift instruction.

3) Block data transfers to and from memory. Unlike regular load/store instructions which operate on a single register, block transfers operate on a list of registers as specified by a 16-bit field. This is a multi-cycle operation which halts subsequent instructions in the pipeline until the operation has completed.

4) Multiply and multiply-accumulate instructions. They require special hardware to implement, and impact data flow because of the 64-bit result generated by a 32 x 32 multiply and the need for the result to pass through the ALU to perform the accumulate.

5) Complex addressing modes. The ISA supports both immediate address offsets from the base address register, and register-shifted offsets. In addition, the ISA allows pre/post indexing and auto increment/decrement addressing modes *[6.3]*.

ARM8 Pipeline

The basic ARM8 pipeline has five stages, as shown in Figure 6.4. However, the *Write* stage is only used by load instructions when writing data to the register file. All other instructions (e.g. data processing, store, branch, coprocessor) complete by the end of the fourth stage at which time they have completed any writes to the register file. The

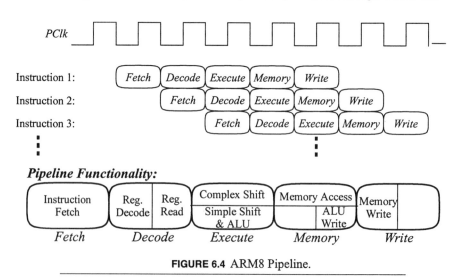

FIGURE 6.4 ARM8 Pipeline.

coprocessor operates lock-step with the ARM8 pipeline, but with a half-cycle delay, which is described in more detail in Section 6.2.5.

Complex instructions will cause the core to iterate on a pipeline stage more than once, forcing all previous pipeline stages to halt. Simple shifts (left-shift by zero, one, two, or three bits) are optimized to complete in the same cycle as any operation requiring the ALU, but complex shifts, which require the use of the barrel shifter, force the core to loop on the *Execute* stage twice - one full cycle for the shift, and another full cycle for the ALU operation. The multiply instructions will cause the core to loop on the *Execute* stage a variable number of cycles, depending on the operands' data. Block data transfers will force the core to loop on the *Memory* stage until all the requisite registers have either been loaded or stored.

ARM8 Data Flow Architecture

Figure 6.5 present a block diagram of the ARM8 and highlights the main sub-blocks of the core's datapath. The prefetch unit fetches instructions from the memory via *RData* and places them in an eight-deep FIFO. The memory address is generated by the program counter (PC) incrementer block and placed on *VAddress*. Whenever there is no load or store pending on the memory bus, and the FIFO is not full, the prefetch unit will fetch more instructions from memory. The datapath fetches instructions from this FIFO in the first (*Fetch*) pipeline stage, and decodes the instruction in the first half of the *Decode* stage.

In the second half of the *Decode* stage, the register operands are read from the register file and placed on *ABus* and/or *BBus*. At the beginning of the *Execute* stage, these busses input data to the multiplier, the write-data pipeline, or the ALU. The multiplier will stall the datapath for two to six additional cycles, depending upon how many of the most-significant source-operand bytes are zero, and place the output product back onto *ABus* and/or *BBus*. The product passes through the ALU to be written back into the register file. The write-data pipeline holds the register's data to be saved for one cycle until the *Memory* stage, at which time the data is then placed on *WData*. Simple ALU operations will complete in one cycle during the *Execute* stage, place the result on the *Result* bus, and write the value back to the register file during the second half of the *Memory* stage.

For loads and stores, the ALU is used to calculate the effective memory address during the *Execute* stage, and the address is sent to the address buffer via the *Result* bus. The address is then placed onto *VAddress* during the *Memory* stage. Stores will complete in this stage by placing the stored data onto *WData*. Loads will be initiated dur-

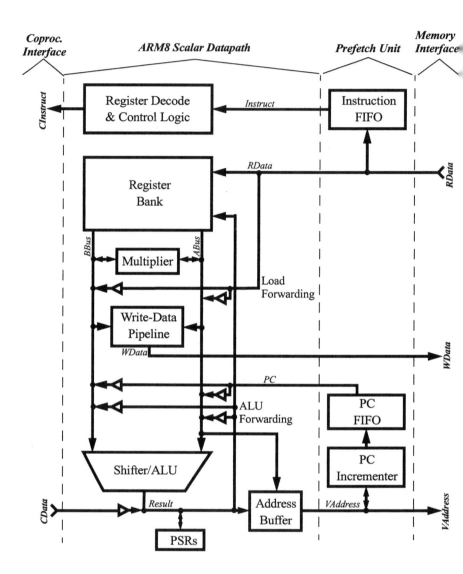

FIGURE 6.5 ARM8 Data Flow Block Diagram.

ing this stage, but the memory data will not be valid until the end of this stage, and written to the register file in the first half of the *Write* stage.

When the PC is used as an operand register, the PC FIFO is used to hold the value until the end of the *Decode* stage, at which time it is placed onto either *ABus* or *BBus*. Writes to the PC are done via *VAddress* and flush all previous instructions in the pipeline, inducing a two to four cycle penalty, depending upon the instruction. At the same time the PC is placed on *VAddress,* to be latched into the PC incrementer block, the instruction at that location is fetched and the prefetch unit begins loading subsequent instructions.

To remove read-after-write (RAW) hazards in the pipeline, there are two sets of data-forwarding paths. The ALU-forwarding path can bypass the *Result* bus to *ABus* and *BBus* when a data processing or effective-address calculation result is used as an operand in one of the next two subsequent instructions. It is either immediately forwarded at the end of the *Execute* stage, or at the end of the *Memory* stage, in which it is simultaneously being written back to the register file. Load instructions do not return their data value to the datapath until the end of the *Memory* stage, so if the very next instruction uses the loaded register as an operand, the datapath must stall for one cycle. To make the data available for the *Execute* stage that coincides with the load's *Write* stage, the load-forwarding path puts the returned data value onto either *ABus* or *BBus* while simultaneously writing the data to the register file.

The 32-bit Process Status Register (PSR), shown in Figure 6.6, contains the condition code flags, the interrupts disable flags, and the mode bits which indicate which mode the ARM8 is operating in (e.g. user, supervisor, interrupt). Writing to the PSR depends upon the current operating mode, and is done via the ALU and the *Result* bus. Reading from the PSR is allowed in any mode, and is done by placing the register value onto the *Result* bus and writing it to the register file. Instructions that set the condition code flags do so at the end of the *Execute* stage. Subsequent conditional instructions proceed as normal through the pipeline, but can be flushed in either the *Decode* or *Execute* stage once the value of the flags are known

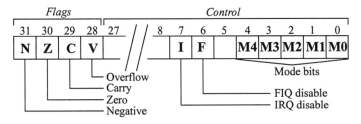

FIGURE 6.6 Process Status Register.

The ARM8 interfaces to the system coprocessor via three busses. The pre-decoded coprocessor instruction is placed onto *CInstruct* in the first half of the *Decode* stage so that it is available on the rising edge of *PClk*. On this edge, the coprocessor enters its half-cycle delayed decode stage. Reads from the coprocessor to the ARM8 are performed via *CData*, which is driven during the coprocessor's execute stage so that data can be written to the ARM8's register file in the second-half of the *Memory* stage. Writes to the coprocessor are done by placing the data value on *VAddress* during the *Memory* stage, which makes the data available to be written into the coprocessor's register file during its memory/write stage. The coprocessor pipeline is described in more detail in Section 6.2.5.

ARM8 Memory Interface

Although a split instruction/data cache structure is generally more energy efficient, the ARM8 memory interface is designed to connect to a unified cache due to legacy system architecture constraints. Since, on average, there is more than one memory read per instruction, due to instruction fetches and loads, the interface is designed to return up to two words per cycle to eliminate this bottleneck. While this would seem to shift the bottleneck to the cache memory's critical path, Section 6.2.2 will demonstrate that by constraining when the cache will return two words, the memory's critical path can be significantly reduced with little or no degradation of processor performance.

The timing of the memory interface is shown in Figure 6.7. Both *VAddress* and the *ARequest* control signal are driven by the ARM8 when *PClk* is high so that they are available to the cache system on the falling edge of *PClk*. If the memory request is a store, then the data word to be written is placed on *WData* when *PClk* is low. The ARM8 always expects to get acknowledgment of the memory access request from the cache controller by the rising edge of *PClk* via the *AResponse* control signal. For instruction fetches and loads, the ARM8 expects the word to be available by the falling edge at the end of the memory access cycle. If two words are requested, it expects the second word to be available on the rising edge of *PClk* following the access cycle.

If the cache system cannot complete the access request in one cycle, it must still return an acknowledgment on *AResponse* when *PClk* is low, then deassert the *Confirm* signal when *PClk* is high. Forcing *Confirm* low stops the clock signal in the core clock control domain. The cache system reasserts *Confirm* when *PClk* is high during the cycle it can complete the request, as illustrated for the fifth access sequence in Figure 6.7. Due to the nature of the memory pipeline, the ARM8 will already have placed the sixth access request onto the interface, but it will stall until the fifth access request has completed.

Processor Architecture

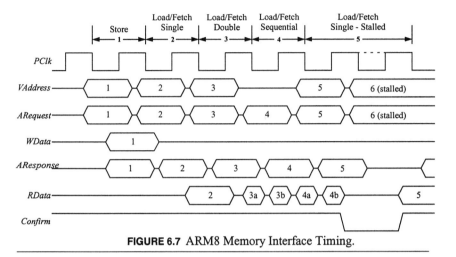

FIGURE 6.7 ARM8 Memory Interface Timing.

The ARM8 encodes in the *ARequest* control signal whether the access request is sequential to the last request of similar type (i.e. instruction fetch, data load, data store), whether an instruction fetch is speculative or not, and whether the data load/store will have more sequential accesses to follow as part of a block transfer instruction. Access requests that are sequential do not need to drive *VAddress* because the cache system can infer the new address by incrementing the previous address. These hints from the ARM8 are utilized to improve the energy efficiency and performance of the cache, which is described in further detail in Section 6.2.2.

Optimizations for Energy Efficiency

The ARM8 RTL behavioral model specified the microarchitecture of the core, and was segmented into 29 sub-blocks. In order to use the model's companion test vector suite, which provided vectors for the complete core as well as the individual sub-blocks, the microarchitecture could not be altered. Generating a new suite requires a tremendous effort, and outweighed the potential improvement in energy efficiency that might be achieved by modifying the core's microarchitecture. Thus, only the physical implementation of the processor core was optimized.

Before starting the design, an effective switched capacitance budget for the core was set. The budget is in capacitance, and not energy consumption, because with DVS the energy will vary with V_{DD}, but the effective switched capacitance will remain roughly constant. A budget was necessary to speed up design time so that only those blocks that significantly contribute to the total core capacitance were energy optimized. The design optimizations utilized the circuit design methodology outlined in Chapter 4.

Previous analysis [6.4] and discussion with ARM (regarding an ARM7 core) indicated that for simple, scalar processor cores, three blocks -- the register file, shifter, and ALU -- contribute more than 50% of the total processor capacitance/cycle. However, the ARM8 has a prefetch unit, whose additional complexity was estimated to reduce the contribution of these blocks to approximately 33%. Energy-efficient test implementations of the three blocks in the target 0.6μm process technology were 50 pF/cycle, which was then multiplied by three to get the budgeted effective switched capacitance of 150 pF/cycle. This budget was believed to be an aggressive, yet achievable goal.

Previous research has demonstrated that for a large enough sample of machine code, there is little variance in the effective switched capacitance per cycle for scalar microprocessors [6.6]. Since the test vectors for each of the sub-blocks originated from machine code run on the entire core to thoroughly test the sub-block, the capacitance/cycle measured for each sub-block executing its own test vectors, when summed for all sub-blocks, should approximately equal the capacitance/cycle measured for the entire core running typical machine code. This critical observation allowed each of the sub-blocks to be optimized in isolation, greatly reducing simulation time, and yielding an overall energy-optimized processor core because individual sub-blocks can be simulated much faster than the entire processor core.

This is validated in Figure 6.8 which compares the measured capacitance/cycle when simulating the entire core (black) versus when simulating an individual block (white). All of the individual sub-blocks compare to within 20%, with the exception of the last sub-block, the multiplier, because it is exercised much more heavily in the test code than in typical machine code. The 20% maximum variation can be reduced to approximately 10% if the global bus capacitance is modeled in the sub-block simulation, which is not the case for the measured data in Figure 6.8.

To further speed up the design time, the schematics were first energy optimized before the time-intensive task of committing them to custom layout. Figure 6.8 compares the measured capacitance/cycle for each sub-block when simulating the schematic and extracted layout netlists. The relative capacitances compare very well, with the extracted netlist yielding slightly higher capacitance due to the inclusion of interconnect capacitance. Only two blocks radically deviate, which are the register file, due to the overestimation of drain capacitance on the bitlines, and the ALU/shifter, due to a large number of busses adding significant interconnect capacitance.

In summary, the bulk of the design for energy optimization occurred while designing the schematics for the various sub-blocks, which could be simulated individually providing fast feedback on the measured capacitance per cycle. For those sub-blocks that were below 3% of the budgeted capacitance (4.5pF), which were a majority (18 of 29 sub-blocks), only obvious energy optimizations were made and the schematics were quickly mapped to layout. This allowed more time to be spent on the nine remaining blocks to be carefully optimized for energy efficiency, making the best use of the design effort and yielding an overall energy-efficient processor core implementation.

Core Energy Breakdown

A breakdown of the processor core energy consumption is shown in Figure 6.10. The numbers were generated from a 25,000 cycle simulation of typical machine code on the extracted layout of the entire core. To ensure that this was a reasonable simulation, it was observed that the energy consumption is within 10% of the final value in less than 10% of the simulation time. The simulation takes into account processor core stalls due to memory system latency since the input vectors were generated from

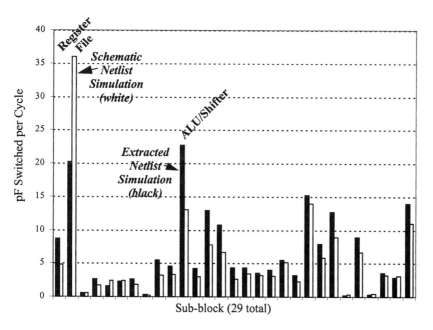

FIGURE 6.8 ARM8 Capacitance of Schematic vs. Extracted Simulation.

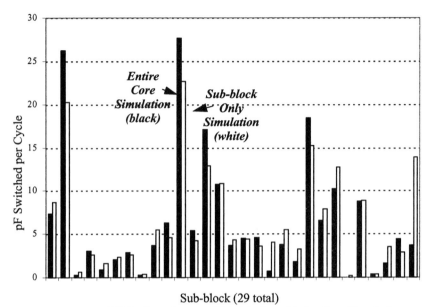

FIGURE 6.9 ARM8 Capacitance of Sub-block vs. Entire Core Simulation.

a full system simulation. However, during this simulation, the processor core is never put into sleep mode, and requires an average effective switched capacitance of 187 pF/cycle while the processor system is active.

The breakdown demonstrates that the scalar core consumes 71% of the total energy, split 60-40% between the custom-layout datapath and the fully synthesized control logic. The prefetch unit consumes 27% of the total energy, split 70-30% between the datapath and control, while the multiplier consumes only 2% of the total energy for typical machine code. Among all sub-blocks, only six of them contribute 59% of the total capacitance/cycle, and it was on these six sub-blocks that a significant fraction of the design time was spent.

The register file, ALU, and shifter combined contribute 54 pF/cycle, validating the assumption in Section , which estimated the capacitance at 50 pF/cycle, and subsequently utilized to derive the capacitance budget for the entire core. These three blocks' capacitance/cycle is only 29% of the total (40% of the scalar datapath/control) which validates the initial assumption that these three blocks would contribute one-third of the total core capacitance budget of 150 pF/cycle.

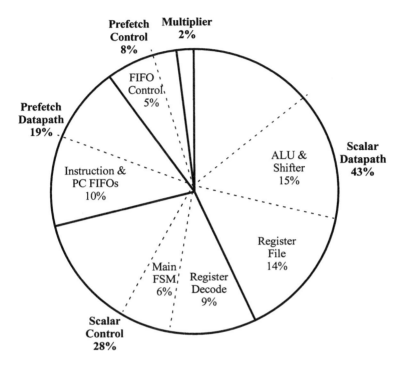

FIGURE 6.10 ARM8 Energy Breakdown (Full Core Simulation).

6.2.2 Cache

For scalar processor cores, the cache typically dominates the total microprocessor energy consumption. However, since there was complete freedom in the design of the cache, this implementation was heavily energy-optimized, yielding a very energy-efficient cache that consumes only about one-half of the core energy consumption. The only constraint on the cache was that it should be unified, and support two memory reads per cycle to accommodate the ARM8 memory interface. The size of the cache was chosen to be 16kB to maximize system energy efficiency (Section 3.3.1), and was solely limited by microprocessor die size.

The cache characteristics and policies were optimized for energy-efficiency as described in Section 3.3, and summarized below:

- Associativity: 32-way. Each 1kB block has a 32 x 23 bit CAM for the tag lookup.
- Line Size: 32-bytes. There are 32 lines per 1kB block.
- Write Policy: Write-back. Main memory is updated only when a dirty cache line is replaced in the cache.
- Write Miss Policy: No write allocate. Write misses are sent directly to external memory, and are not placed within the cache.
- Replacement Policy: Round-robin. Successive lines in the 1kB block are chosen for replacement upon a cache miss. A line will not be replaced until the 33rd cache miss to a particular block.
- Double Reads: Two words may be returned in a single cycle if the address LSB is 0. If two words are requested and the LSB is 1, only one word is returned. After the odd-address read, the ARM8 prefetch unit will become even-address aligned allowing subsequent double reads.
- Instruction Buffer: Each 1kB block has an implicit instruction buffer, though only one is active at a time. Consecutive instructions that do not cross a cache-line boundary can be made without activating the CAM, reducing energy consumption by 50%.

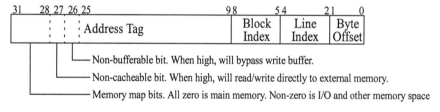

FIGURE 6.11 Address Space Breakdown.

Figure 6.11 shows the how the 32-bit address space is utilized. The cache is physically addressed, eliminating the need for a translation look-aside buffer (TLB). For embedded applications, a TLB is not particularly useful since embedded systems generally do not have a larger, secondary storage unit (e.g. disk drive) which requires a TLB to map it onto the smaller physical memory. Since a TLB also provides separate address spaces to prevent memory conflicts, the lack of one in this system forces the operating system and/or programmer to ensure no memory conflicts exist.

Cache Memory Array

As a compromise between energy consumption and silicon area (as described in Section 3.3.2) the basic block size was set to 1kB, and replicated 16 times to form the cache memory array as shown in Figure 6.12. Due to the large size of the cache, which fills approximately 60% of the chip, careful attention had to be paid to routing of the busses, control signals, and power lines.

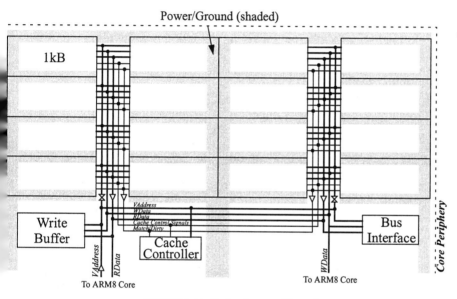

FIGURE 6.12 Cache System Floorplan

The shaded areas indicate where power and ground are routed, which use *Metal2* & *Metal3*. *Metal1* is used to connect up the bypass capacitance sitting under the power routes, which totals 11.7nF for the entire array. The left, right, and top sides of the cache abut the pad ring, providing low impedance from the power pins distributed around the periphery to the entire cache memory.

The switching activity of the three cache busses was analyzed to calculate the total effective switched capacitance per cycle, as shown in Table 6.1. The capacitance on *WData* is negligible, due to the low ratio of processor writes to reads, and the high correlation in the data being written. The capacitance on *VAddress* and *RData*, however, is significant (2.2% and 8.1%, respectively, of the total chip's capacitance/cycle budget). By inserting a bi-directional switch on *VAddress* and a uni-directional switch on *RData*, only half of the cache toggles per cycle, reducing the capacitance/cycle by 2.3pF (-0.8%) and 6.5pF (-2.2%), respectively. By buffering up the signal on either half of the cache, the capacitance that each block's outputs have to drive is reduced 60-75%, such that much smaller drivers can be used. The speed-up of driving less capacitance offsets the two-gate delays contributed by the insertion of the switches. To lighten the load on the *Match* and *Dirty* output signals, they too have unidirectional switches, reducing each block's output capacitance that it has to drive by 60%.

TABLE 6.1 Cache Bus Switching Activity and Effective Switched Capacitance.

Bus	Toggles (0→1) per cycle	Capacitance per Bit			Total Bus Capacitance per Cycle
		Half Cache	Global	Total	
VAddress	1.95	1200 fF	900 fF	3300 fF	6.4 pF
Wdata	0.15	600 fF	840 fF	2040 fF	0.3 pF
Rdata	6.50	1000 fF	1750 fF	3750 fF	24.4 pF

Thus, through simple high-level simulation and energy analysis, more than 3% of the total microprocessor's energy consumption was reduced with the addition of these simple switches.

Cache 1kB Macro

The 1kB macro builds upon a previous energy-efficient SRAM design *[6.5]*, which was ported from a 2-level metal process to a 3-level metal process. The additional metal layer was used to provide much better power distribution and reduce the capacitance on the bitline.

The architecture of the macro is shown in Figure 6.13. On the left side is the CAM array which contains the 23-bit address tags for the 32 cache lines. Upon a cache read, *prematch[31:0]* is precharged and the 23-bit address tag is passed into the CAM array, described in further detail in Section 4.2.2. If the *n-th* tag in the CAM array matches, the *prematch[n]* signal remains asserted while all the other bus signals are pulled low. The CAM state latches block contains the valid state bit for each of the 32 CAM tag addresses. The asserted *prematch[n]* signal is AND-ed with its corresponding valid state bit to indicate whether a valid match exists, and if so, the signal *match[n]* gets asserted as well as the global *Match* signal which is sent to the cache controller. This indicates the desired cache line is present in the block.

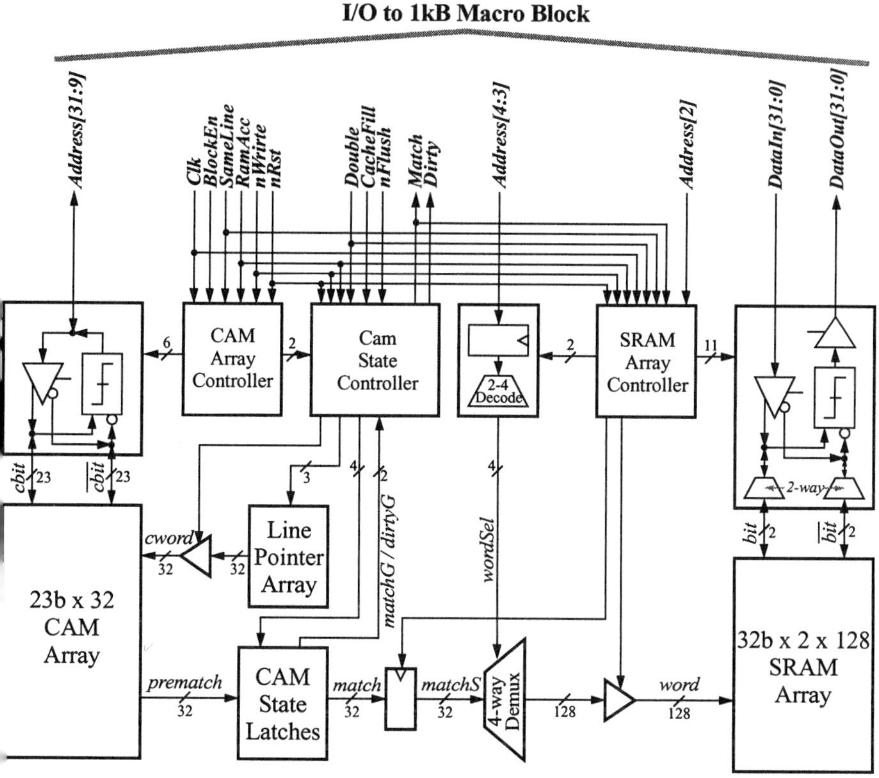

Processor Architecture

FIGURE 6.13 Cache Memory 1kB Macro Architecture.

The *match[31:0]* bus gets latched for subsequent sequential cache reads. The *matchS[31:0]* bus is demultiplexed by *Address[4:3]* to select the desired word-pair of the cache line, and drives the corresponding *word[m]* signal into the SRAM array. For a cache read, the two 32-bit data words corresponding to *word[m]* are read from the array, and *Address[2]* drives the column demultiplexer to select which one of the two words to place onto *DataOut[31:0]*. A cache write will take the data off *DataIn[31:0]* and write it to the desired location in the SRAM array. The schematic of the SRAM array cell, column decoder, and sense-amp is described in Section 4.2.1.

During a cache read, if the global *Match* signal remains low, indicating the cache line is not present in the block, then the cache controller looks at the global *Dirty* signal. *Dirty* is set at the end of the match operation if the next location to be replaced in the CAM has been written to in the cache and needs to be updated in main memory before replacing. The next location is determined by the line pointer array block, which is a circular chain of 32 latches, and gets rotated when a new cache line is written to the macro block. If the cache line is dirty, then the cache controller reads the address tag out of the CAM array, and then its corresponding cache line, which is then written to main memory. To place new data into the cache, the cache controller first writes the new address tag to the CAM array, and then subsequently, the eight data words corresponding to this cache line.

Because *matchS[31:0]* latches the last cache line that matched, subsequent cache accesses, which are sequential and do not wrap to the next cache line, do not need to access the CAM. Instead, the cache controller increments the cache-line index bits *Address[4:2]* appropriately, and *matchS[31:0]* drives the desired *word[m]* line to access the SRAM array.

Cache Controller

The cache controller state diagram, shown in Figure 6.14, contains 30 unique states. It is the central controller for the entire cache system, driving not only the 16kB cache memory, but also routing data to the write buffer and to/from the external interface. The bulk of the states are required to manage the cache memory, which include writing dirty cache lines to main memory, reading in new cache lines, flushing the cache memory, and performing read-modify-write operations to support the ARM8's byte and half-word operations.

To demonstrate the timing of the cache system, a timing diagram for a double-read to the cache is shown in Figure 6.15. The ARM8's address and control signals arrive at the cache controller during the *Phi2* clock phase. The cache controller must set both the correct block enable signal and the macro block control signals by the end of the *Phi2* clock phase, so that they are stable when the CAM is accessed in the *Phi1* clock phase. The cache controller must always return a response to the ARM8 by the end of *Phi1*. If *Match* remains high, then the cache line was found in the macro block, and in the subsequent two clock phases, data is returned to the ARM8 via *RData*. If *Match* goes low, then the cache controller will lower *Confirm* (not shown) in the next *Phi2*, which will gate the clock to the ARM8 core while the cache system fetches the desired word. In the meantime, the cache controller loads in the desired cache line from main memory, and if *Dirty* was also low, then it writes out the old cache line back to main memory.

Processor Architecture

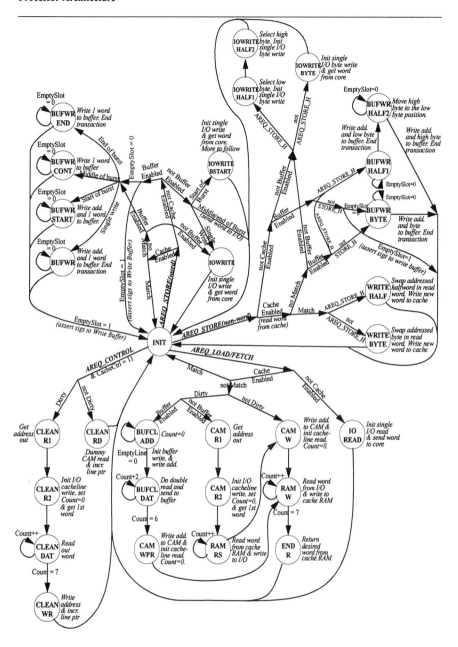

FIGURE 0.14 Cache Controller State Diagram.

Microprocessor and Memory IC's

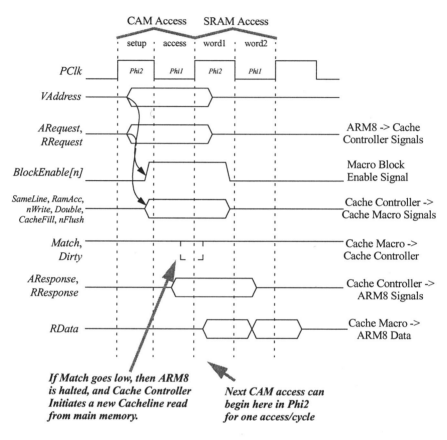

FIGURE 6.15 Cache Memory Timing for a Cache Double-Read Hit.

Cache Design Optimizations

The cache controller's control signals to the write buffer, bus interface, and the ARM8 core are dependent upon whether the *Match* signal, which is output by the activated cache macro block, is high or low. The signal is not available until late in *Phi1*, and created a critical path for generating these control signals, which must be available at the beginning of the next *Phi2*. All possible cache accesses were analyzed with the C simulator and categorized as either common cases or rare cases as shown in Table 6.2 . To reduce the critical path, an extra cycle delay was added to the state

192 *Energy Efficient Microprocessor Design*

machine for the rare cases, in order to reduce the loading on *Match* and speed up the critical path.

TABLE 6.2 Categorization of Cache Access Types.

Common cases requiring optimization:	Rare cases which could be slowed down:
1. Cache read hit (single & burst)	1. Non-cacheable burst reads
2. Non-cacheable single read	2. Cacheable/ bufferable write miss (single & burst)
3. Cache read miss (single & burst)	3. Cacheable/ non-bufferable write miss (single & burst)
4. Cache write hit (single & burst)	
5. Non-cacheable/bufferable write (single & burst)	
6. Non-cacheable/non-bufferable write (single & burst)	

Byte and half-word reads are rotated by the core. Byte and half-word stores must be rotated by the memory system. Since the cache controller only consists of standard cells, the datapath logic to do this resides in the bus interface, where the *RData* and *WData* busses are readily available. However, this datapath logic is directly controlled by the cache controller. Writes to memory locations present in the cache memory require one stall cycle so that a read-modify-write can take place, as shown in the timing diagram in Figure 6.16. The original data word is read from the cache memory, and latched from *RData* onto *RDataT2*. The byte or half-word to be written is latched off of *WData* onto *WDataT1*, then merged with the saved data on *RDataT2* and placed back onto *WData* where it can be written to the cache memory. Write misses are sent as byte writes to either the write buffer or the bus interface. Because the external SRAM and I/O can only operate on words and bytes, two cycles are required for half-word write misses, in which *WDataT1* is used for temporary storage, in order to split the half-word into two byte writes. When *Confirm* goes low, the ARM8 core is stalled for either one or two cycles depending upon whether it is a byte or half-word write.

Cache Energy Breakdown

As shown in Figure 6.17, half of the energy consumed by the cache occurs in the SRAM component of the cache memory. For a single cache access in which the CAM is activated, the CAM consumes 60% of the energy consumed by the SRAM. But with the virtual instructions buffers, activation of the CAM is suppressed for sequen-

Microprocessor and Memory IC's

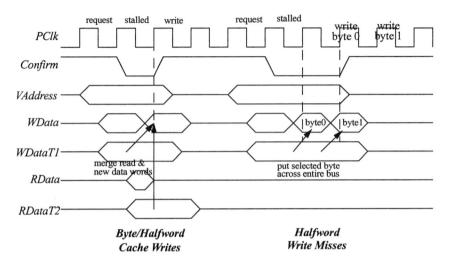

FIGURE 6.16 Timing for Non-word Writes to the Cache and Main Memory.

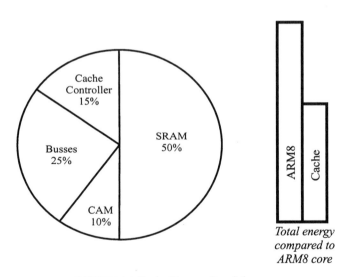

FIGURE 6.17 Cache Energy Breakdown.

Energy Efficient Microprocessor Design

tial instruction fetches, reducing its average energy consumption to 20% of that of the SRAM. The busses consume 25% of the total cache energy, and includes the energy consumed by the address buffer. Finally, the cache controller consumes 15% of the total cache energy. On a cycle-by-cycle comparison, the cache consumes 63% of that consumed by the ARM8 core.

6.2.3 Write Buffer

Since the external bus multiplexes address and data onto the same bus, the write buffer stores both the address and data in a single register file, as shown in Figure 6.18. The multiplexed architecture allows either one cache line and one store,

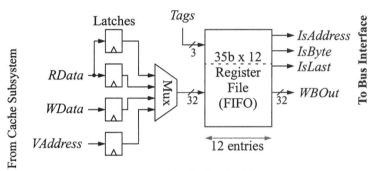

FIGURE 6.18 Write Buffer Architecture.

or up to six single-word stores. In addition, the buffer can store a variable-number of multiple words per single address for the Store Multiple (STM) instruction. If the STM words cross over a cache-line boundary, the beginning address of the second cache line is placed into the buffer to align the external memory access on a cache-line boundary. This is required to ensure that the store does not cross over multiple external SRAMs, which the bus interface cannot support. Simulation demonstrated that given the system architecture, any more than twelve buffer entries yields negligible performance improvement.

The register file is 35 bits wide. The additional three bits are tags used to indicate if the entry is an address (*IsAddress*), if it is the last data word (*IsLast*) or if it is a byte-wide store (*IsByte*). The three busses of the cache subsystem (*VAddress/Wdata/Rdata*) are latched and multiplexed going into the register file. The latches are required to provide enough setup and hold time for the register file, with two latches required for

RData because when reading out a line from the cache memory, two words are returned per cycle. The address/data words are placed onto *WBOut* and sent to the bus interface, under its control.

An out-of-order write buffer requires hardware to compare the address of a pending read to all the addresses stored in the write buffer to ensure memory consistency. The buffer control was significantly simplified by enforcing all external memory accesses to be in order; before any external read request, the write buffer is first flushed out. The exception to this rule is cache-line reloads, which are guaranteed not to have the same address between the cache line being written out and the new line being read in. Providing this exception reduces the latency to complete a cache-line reload by a factor of two.

The input to the write buffer is controlled by the cache controller via four signals, shown at the bottom of Figure 6.19. The *LoadWord* signal is utilized to enable the write buffer, while the other three signals are decoded to determine which bus to latch (*LoadDirect*), which word is the address (*LoadAdd*), and which data word is the last one (*LoadLast*). The timing on the input busses was dictated by the ARM8 memory interface, and the timing of the cache memory.

Energy Consumption

For single writes (STR), the effective switched capacitance is 63 pF/word (from simulation on extracted layout), and 26 pF/word for multi-word stores (STM). The only instructions which use the write buffer are non-cacheable stores and stores that take a cache miss. Since stores are approximately 10% of the instruction mix, the write buffer contributes 1.3 pF/cycle, on average. Read cache misses may also enable the write buffer, but only if the cache line is dirty, and switch 110 pF/cacheline. However, this condition occurs well under 1% of the time, so the overall contribution is less than 1 pF/cycle, on average. Thus, the write buffer consumes less than 1% of the total processor chip energy consumption.

6.2.4 Bus Interface

The primary responsibility of the bus interface is to connect the processor clock (*PClk*) and bus clock (*MClk*) domains. The bus interface also includes the components to enable byte and halfword writes, as shown in Figure 6.19. Reads from external memory typically take many cycles to complete, and will stall the ARM8 core and/or cache system until the external access has completed. Since the prototype processor is an in-order machine, there is no reason to provide buffering for reads to

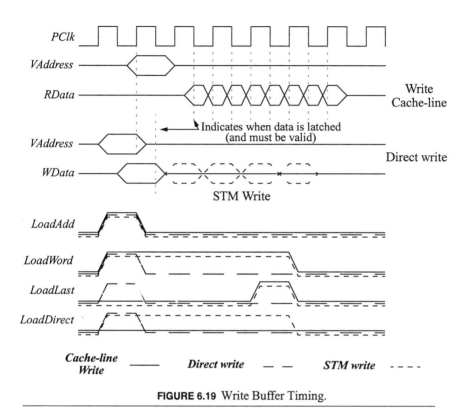

FIGURE 6.19 Write Buffer Timing.

allow the core to continue operating, since it must wait for the pending word to continue. With a separate write buffer, there is no need to provide additional buffering within the bus interface, such that the bus interface complexity is reduced to a four-to-one multiplexer, four registers, and enabled buffers to drive the cache-system busses.

The bus interface talks to the cache controller, and the core via the controller, on one side of the interface running at the processor clock speed (*PClk*). The other side of the interface communicates with external memory at the processor bus clock speed (*MClk*). The *MClk* speed is programmable via the system coprocessor, and can operate at a 2x, 4x, or 8x multiple of *PClk*. Initially there was a 1x option, but the additional hardware to support this was not warranted given the marginal performance improvement achieved.

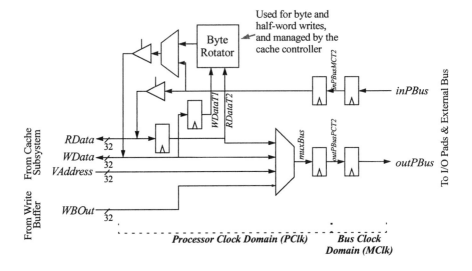

FIGURE 6.20 Bus Interface Architecture.

The state machine controlling the bus interface is relatively simple, as shown in Figure 6.21. If the state machine is idling, it services the write buffer if it is not empty. Otherwise, it services I/O requests from the cache controller. Maintaining this order ensures memory consistency. Otherwise, read requests from the cache controller would have to be matched against pending writes in the write buffer, at the expense of significant hardware complexity. Simulation demonstrated little performance degradation by enforcing this order to eliminate the extra hardware. The only instance when the cache controller takes priority over the write buffer is for a dirty cache-line load, in which it is guaranteed that the outgoing cache line is not the same memory location as the incoming cache line.

All bus transactions must complete before moving onto a new transaction. The state machine operates at *PClk* speed, and sends signals to the bus-side logic via a simple handshake scheme that is independent of the phase difference between *PClk* and *MClk*. If the *PReq* signal is asserted by the I/O chip, indicating a pending direct memory access (DMA) request, the state machine hands off control of the bus after completing all outstanding bus requests. The I/O chip can then directly access main memory. The processor core can continue to run, but if it attempts to access the bus,

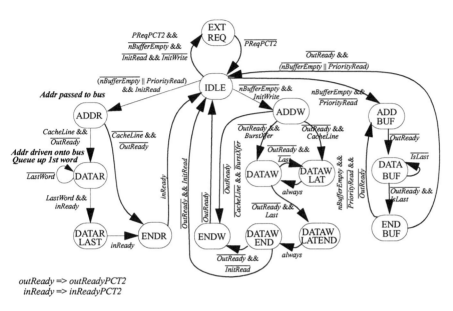

FIGURE 6.21 Bus Interface State Machine.

or attempts to write to a full write buffer, then the core will stall until the I/O chip releases ownership of the processor bus.

Clock Interfacing

MClk is derived from *PClk* using a selectable frequency divider consisting of three flip-flops and a multiplexer, which introduces some phase shift. The *MClk* signal actually used is a buffered version of the signal off the external clock pad, which ensures that the processor chip, memory chips, and I/O chip all operate with an *MClk* that has minimum relative phase shift between the chips. This improves the robustness of the signal timing on the external processor bus. This buffered *MClk* signal introduces additional phase shift with respect to *PClk*, but a self-timed handshake scheme allows proper operation of the bus interface independent of this phase shift, as well as V_{DD}.

Core->Bus: The state machine changes state in *Phi2* of *PClk*, and all signals going to the external bus are derived from the machine state and other *Phi2* signals. On the falling edge of *PClk*, the control signals are latched, and the signal *outReady* is asserted via an $\overline{\text{RS}}$ latch, with some delay. The *outReady* signal is then latched when

Microprocessor and Memory IC's

MClk is low, generating *outReadyMCT1*, to ensure a stable signal when *MClk* is high. Upon the next rising edge of *MClk*, if *outReadyMCT1* is high, the output data is latched, placed on *outPBus*, and sent directly to the processor bus pads. At the same time, the bus-side logic drives *outReady* low via the same $\overline{RS}$ latch, and *outReady* is latched on the rising edge of *PClk* to generate the signal *outReadyPCT2*. This signal is used by the state machine to either wait, or pass new data to the processor bus. When *outReady* and *MClk* are coincident in time, if the bus-side logic detects it is high, the additional delay generating *outReady* will ensure the processor-side data is valid. Otherwise, *outReady* will not be detected until the subsequent rising edge of *MClk*. The signal *outReadyPCT2* can stay high for up to one *PClk* cycle after the rising edge of *MClk*, but since the lowest clock ratio is 2x, the core-side logic will set up the next data element in the second cycle, and be ready for the next slot on the processor bus. Hence, the state machine at all times will be able to maintain maximum throughput on the external processor bus. The timing for when the edges are coincident are demonstrated in Figure 6.22.

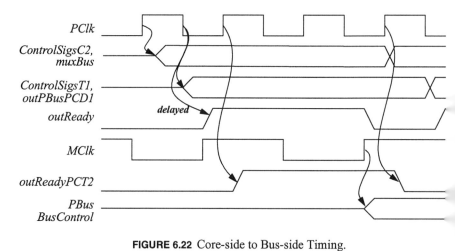

FIGURE 6.22 Core-side to Bus-side Timing.

Bus->Core: This case is similar to the previous one, in which all outgoing processor bus requests get latched on the falling edge of *PClk*, and latched on the next rising edge of *MClk* when *outReadyMCT1* is high. There is an additional signal that is sent to tell the bus-side logic to latch the external processor bus on the falling edge of *MClk*. When this occurs, the signal *inReady* gets asserted via another $\overline{RS}$ latch, which is latched on the next rising edge of *PClk* to generate the signal *inReadyPCT2*. Again, additional delay is placed on the *inReady* signal to give the latched data enough time to settle. The signal *inreadyPCT2* is then used by the state machine to latch the data taken of the *inPBus* bus, and send it off to the cache or processor core.

It is possible, if the *outReady* signal is exactly coincident with the rising edge of *MClk*, for it to be detected some cycles, and not for others. This has the potential of placing a bubble on the bus if a missed cycle follows a caught cycle, leading to an invalid bus operation and possible system failure. To prevent this, an additional signal can be utilized to shift *MClk* an additional eight gate delays. This will eliminate the coincidence, thereby eliminating the coincidence. Fortunately, this feature was not required for correct operation of the prototype processor system.

Energy Consumption

Due to its simplicity and infrequent use, the bus interface has very little energy consumption, on the order of 1-2% of the total processor energy consumption. However, this does not include the bus drivers located in the chips pads, which consume considerably more energy due to the large capacitance on the external bus, on the order of 5-10% of the total system energy consumption.

6.2.5 System Coprocessor

The system coprocessor is a standard component of an ARM-based microprocessor system, and is commonly found in some form in most other microprocessors as well. It is responsible for system-level control functionality which is independent of the processor core, as well as configuring the specifics of the processor core.

The coprocessor operates lock-step with the ARM8 pipeline, but with a half-cycle delay as shown in Figure 6.23. For example, the coprocessor's *CDecode* stage starts on the rising edge of *PClk* in the middle of the core's *Decode* stage. All instruction fetching is performed by the core, so the coprocessor has no fetch stage. There is also no memory stage since the coprocessor cannot directly access memory.

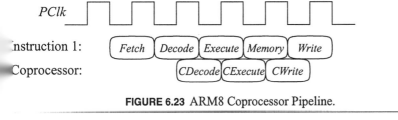

FIGURE 6.23 ARM8 Coprocessor Pipeline.

The ARM8 interfaces to the coprocessor via three busses. The pre-decoded coprocessor instruction is placed onto the *CInstruct* bus in the first half of the processor's *Decode* stage so that it is available on the rising edge of the clock. On this edge, the coprocessor enters the *CDecode* stage in which the coprocessor instruction is decoded. The coprocessor then proceeds with its three-stage pipeline, as shown in Figure 6.23. Reads from the coprocessor to the ARM8 are performed via the *CData* bus which is driven during the *CExecute* stage so that data can be written to the ARM8's register file in the second-half of the *Memory* stage. Writes to the coprocessor are done by placing the data value on *VAddress* during the *Memory* stage, which makes the data available to be written into the coprocessor's register file during the *CWrite* stage.

TABLE 6.3 System Coprocessor Register Summary

Reg#	Coprocessor 13	Coprocessor 14	Coprocessor 15
0	Access Cycle Count (RO)	Real Time Counter Low (RO)	MMU ID (RO)
1	Idle Cycle Count (RO)	Real Time Counter High (RO)	System Control
2	Sleep Cycle Count (RO)	Timer Interrupt	*not used*
3	Wait Cycles (RO)	Interrupt Suspend	
4	Hit Count (RO)	Internal Dynamic Clock Speed	
5	Cached Miss Count (RO)	External Pin Control	
6	Cache Writeback Count (RO)	Hw. Control Tweaks	
7	Uncached Access Count (RO)	Instruction Count (RO)	Cache Operations (WO)

The system coprocessor is comprised of various counters and registers containing special state variables. While the block appears logically like a register file, it could not be implemented as such due to the heterogeneity of the registers; some are read-only counters while other are read-write registers. Also, on many of the logical read-write registers, many of the bits are hard-coded to zero. The implementation used a

Processor Architecture

shared bus architecture, with separate input and output ports. Table 6.3 lists all the registers of the system coprocessor, which is logically organized as three separate coprocessors.

Coprocessor 13

This logical coprocessor only consists of read-only counters. Four of the counters (register 0-3) are used to monitor processor operation. One counter records cycles that the processor core is making a memory request (Access); another tracks when the core is active, but has no memory request (Idle); a third tracks when the processor is asleep (Sleep); and the last one tracks when the core is stalled waiting on the external bus to complete a transaction (Wait). Another four counters (register 4-7) monitor cache operation by recording the number of cache accesses that are hits, misses, dirty cache-line writebacks, and uncached accesses. These eight counters can be utilized by the operating system to adjust processor performance depending upon how the processor is being utilized. For example, if the processor spends a significant amount of time stalled, then processor speed can be reduced because the performance bottleneck is in accessing I/O data.

Coprocessor 14

Registers 1 and 0 are read-only, and form a 64-bit real-time counter whose value is in microseconds. When writing to register 2, any pending timer interrupt is cleared, and a new timer value is set, also in microseconds. When reading this register, the next timer event is returned. The processor enters sleep mode when register 3 is written to. The processor will remain idle until the next interrupt occurs, either due to an external event or due to the timer. Reading register 3 returns the current state of the interrupt lines as indicated in Table 6.4 . The *EnIRQ* bit as specified indicates a pending *IRQ* request from an external source, while the *nTIQ* line indicates a pending timer interrupt from the internal timer. The *EnIRQ* and *nTIQ* lines are merged into a single signal, *nIRQ*, which is then sent to the ARM8 core.

TABLE 6.4 Interrupt Information Bitmap. (CP14R3)

31-10	9	8	7	6	5-0
x	nTIQ	EnIRQ	nIRQ	nFIQ	x

A write to register 5 sets two target internal dynamic clock speeds: one for normal operation, and one for interrupts as shown in Table 6.5 . The special interrupt clock speed can be enabled/disabled with a separate control bit for both *IRQ* and *FIQ* in C15R1. Upon writing to this register with no pending interrupts, the desired clock value is sent to the regulation system via the regulator interface. When either an *FIQ* or *IRQ* arrives (and the corresponding mask bit is enabled in C15R1), the interrupt clock speed is sent to the regulation system. Also, upon de-assertion of the interrupt, the normal clock rate is sent to the regulation system. Reading this register returns the current sampled processor speed, which is not necessarily the same value written. This allows the operating system to get feedback from the voltage converter loop to ensure that it is delivering the target frequency.

TABLE 6.5 Clock Speed Write Bitmap. (CP14R4)

31-15	14-8	7	6-0
ignored	interrupt clock speed	ignored	normal clock speed

The lower three bits of register 5 controls the state of four external pins. There is no other effect of writing to this register, and thus it is the recommended register to use when NULL coprocessor write operations are required. A read from this register returns the last value written. Register 6 controls both the external bus clock ratio (bits 6:5), and the fine-tuning for the VCO (bits 4:0), which is described further in Section 6.2.6. The bus clock ratio can be set to 2x (10 or 11), 4x (01), or 8x (00). Register 7 is a read-only register which maintains a count of the number of instructions executed since processor start-up.

Coprocessor 15

Coprocessor 15 contains standard register definitions in ARM implementations *[6.1]*. However, only those registers that pertain to the prototype system architecture were included; for example the registers that control a translation look-aside buffer (TLB) were not implemented since a TLB was not used.

Register 1 is the system configuration register, whose 15 standard bit mappings are described in Table 6.6 . Those bits which apply to the prototype are in **bold**. The **C**, and **W** bits effect the function of the cache memory system. The **B** and **Z** bits are fed back into the processor core to alter core functionality. The **A** bit is sent to the cache controller to alter response to non-aligned memory accesses. All others bits are read as 0 or 1, and are unalterable. Writing to register 7 will flush, or clean, cache blocks. The Flush operation will invalidate the entire cache. The Clean operation writes out data at the specified address if it is dirty. The entire cache can be cleaned by stepping through all 512 cachelines. These operations also require subsequent writes to the

NULL coprocessor register (CP14R5) to work properly with the cache system. The code sequences are shown in Table 6.7.

TABLE 6.6 System Configuration Register Bitmap. (CP15R1)

	31-15	14	13	12	11	10	9	8	7	6	5	4	3	2	1	0
Name	not used	IE	FE	I	Z	F	R	S	B	L	D	P	W	C	A	M
Initial Value	0.....0	0	0	0	0	0	0	0	0	0	1	1	0	0	0	1
Description	A: Alignment Fault Enable C: Cache Enable W: Write-buffer Enable B: Big Endian Select (else Little Endian) Z: Branch Prediction Enable FE: Enable different clock speed for FIQ (set with C14R4). IE: Enable different clock speed for IRQ (set with C14R4).															

TABLE 6.7 Cache Control Operations. (CP15R7)

Function	opcode_2 value	CRm value	Data	Instruction
Flush ID cache(s)	000	0111	SBZ	MCR p15, 0, X, c7, c7, 0
				MCR p14, 0, X, c5, X
Clean ID single entry	001	1011	VA	MCR p15, 0, Rd, c7, c11, 1
				MCR p14, 0, Rd, c5, X
(SBZ = Should Be Zero, VA = Virtual Address, X = don't care)				

Regulator Interface

In the prototype processor, the system coprocessor is also responsible for interfacing with the separate regulator chip, when the conditions for changing the processor frequency occur (Chapter 8). The interface is synchronized to the regulator chip with the 4 MHz clock signal, and transmits the new seven-bit digital frequency value serially, in order to reduce the pins required. Once the regulator has received the new frequency value, and begins adjusting V_{DD} and the clock frequency accordingly, further frequency change requests must be blocked until the regulator has reached steady-state. However, it is not necessary to do this on the processor. On the DVS chip, new request are denied as long as the internal *Track* signal is high, which indicates that the converter is currently changing V_{DD}, so there is no need for flow control from the converter chip back to the processor. Hence, the only time new requests will be blocked by the interface is when there is a currently pending transaction being serially transmitted

Energy Consumption

There are three different processor operating conditions to consider when analyzing this block's energy consumption: Active (processor is active), Wait (processor is stalled on a memory access), and Sleep. The first is not critical, since the energy consumed by the processor core will dwarf that consumed by the coprocessor. The second is not critical either, due to energy consumption in the cache subsystem which dominates any energy consumed by the coprocessor, and because the energy will be the same between the Sleep and Wait modes. The Sleep mode is most critical since this is the lowest energy mode, with energy consumption dominated by the coprocessor and the global clock distribution, as shown in Table 6.8 .

TABLE 6.8 Estimated Processor Capacitance/cycle by Operating Condition.

Mode	Coprocessor	ARM8	Cache System
Active	6.3 pF	200 pF	125 pF
Wait	3 pF	4 pF	40 pF
Sleep		4 pF	3 pF

Thus, the system coprocessor's circuits which are always active (e.g. real-time counters, interrupt controller, etc.) were optimized to minimize their energy consumption, which was reduced to 30% of the total energy consumed by the processor while in Sleep mode.

6.2.6 VCO

To accommodate process variation over the die, as well as simulation error, the oscillator was designed to be programmable from 50% to 150% of nominal frequency with five bits of control. The frequency control is designed to be glitch-free so that it can be programmed via software through a register in the coprocessor (CP14R6).

The basic oscillator architecture, shown in Figure 6.24, consists of five binary-weighted delay blocks, plus a return path to close the loop. Each of the delay blocks has both a slow and fast path which is selected by the *ctrl[n]* signal. A new value for this signal may be loaded when the *trig1* signal transitions low-to-high. By ensuring that the pass gates in the basic block have switched by the time *trig2* transitions low-to-high, the oscillator will change frequency glitch-free. At system start-up, the VCO operates in its slowest mode (e.g. highest voltage for a fixed frequency) to ensure proper operation. In the initial boot sequence, the operating system can measure how fast the VCO can be operated at, and set it accordingly.

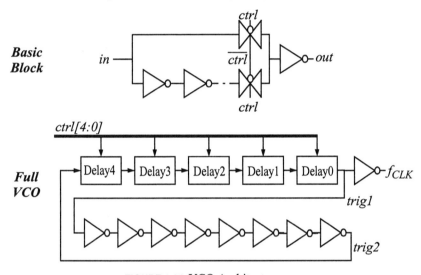

FIGURE 6.24 VCO Architecture.

The hardware was stepped from 5 MHz to 80 MHz in 5 MHz increments, and at each step, the ring oscillator's control bits were decreased until processor failure. Decreasing the control bits had the effect of decreasing supply voltage, since the converter loop maintains constant clock frequency. The minimum control setting to prevent processor failure was exactly the setting for nominal frequency at all frequency values, with the exception at 5 MHz, at which speed the control could be decreased by one LSB from nominal. This demonstrates that the critical paths of a CMOS processor do track extremely well over a wide range of voltage.

6.2.7 Packaging and Chip-Level Design Issues

The microprocessor die was placed into a 132-pin QFP package. There are 77 signal pins, with 56 pins required for the processor system bus including 17 pre-decoded chip-enable signals for the memory chips (*CE[15:0]*) and the interface chip (*IOCE*). Thus, no external decoding circuitry is required, as the processor can be internally configured for 1-16 32kB, 64kB, or 128kB memory chips. Additional signal pins include four pins for the regulator chip interface, two pins for the external interrupt lines from the I/O sub-system, one pin for an initial reset by the regulator, one pin for an external reset signal, and one pin for the reference clock signal used for the internal counters. Twelve more pins are used to provide debug support.

There are 55 power pins, with 28 for ground, 16 for the variable processor core voltage (V_{DD}), ten for the variable I/O circuit voltage (V_{DDIO}), and one for the battery voltage (V_{BAT}). Although separate ground lines for the core and I/O circuits would have been preferable, in order to isolate the core from the noisy I/O circuits, the low-impedance substrate in this process makes this unfeasible. The battery voltage is strictly used to provide electro-static discharge (ESD) protection on those input pins whose signal level is V_{BAT}.

Pad Design

For debugging purposes, the I/O pads were designed to operate at a different voltage than the core, so that they could be left at fixed voltage while the internal core voltage was varied. Thus, all output & input pads support level shifting, with the exception that the four signals which connect to the regulator chip must always be at the nominal core voltage (V_{DD}). The schematic for the level-converting I/O pad is shown in Figure 6.25.

For an outgoing signal, the enabled cross-coupled loads on the complementary NMOS gates provides level conversion from V_{DD} to V_{DDIO}. The ratio of NMOS to PMOS width is dictated by the maximum possible range of voltage conversion. The level conversion was designed to operate from 1V to 3.3V, and with an effective W_P of 3μm, the required W_N was 25μm. Simulation demonstrates that this will correctly operate for V_{DD} as low as 950mV. The *enOut* signal must range from 0V to V_{DDIO}, so the enable signal generated by the core passes through its own level-converter before driving the pads.

The target load capacitance on the output is 50pF, which is sufficient to drive ten chips (4pF each) and four inches of a PCB trace (2.5 pF/inch). Signals must be trans-

Processor Architecture

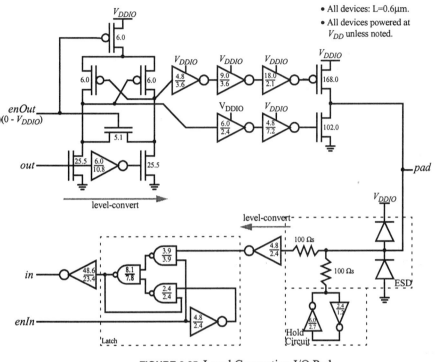

FIGURE 6.25 Level Converting I/O Pad.

mitted within one-half of an *MClk* cycle since they change on the rising edge and are latched on the falling edge. The target delay through the pad is one-quarter of a cycle, allowing another one-quarter cycle of margin before the signal arrives at the other chip. The target rise/fall time is also one-quarter cycle in order to reduce current draw. At 50pF and V_{DDIO} = 3.3V, this corresponds to a peak current of 33mA per signal. The size of the inverters driving the output MOSFET's was dictated by ground and power bounce concerns, and discussed further in Section .

A feedback device was added to the tri-stated outputs to hold state while V_{DDIO} varies. In the prototype system, V_{DDIO} varies at most by 0.2 V/µs, and the output tracks V_{DDIO} to within 30mV. The hold circuit adds negligible delay, and increases the energy consumption within each pad by only 1%.

The ESD protection is comprised of 500 µm² diodes to ground and V_{DDIO}, which is the recommended size according to the process manual *[6.7]*. The diodes provide the primary and only ESD protection, and were validated by simulating the human-body model (HBM) for discharge, which generates a 2kV charge pulse *[6.7]*. The current pulse peaks at 1.33A, and with 5Ωs of series resistance, the voltage rises to 9.1V, which is under the failure limits. To prevent the ESD diodes from turning on, ground and power bounce must be limited to less than 0.5V. Both input nodes have a 100 Ω poly resistor for isolation to prevent gate-oxide breakdown.

The input level conversion is a simple inverter powered at V_{DD}. A latch is used to maintain logic state at the output of the pad, and the output inverter is sized to drive a 1pF load.

Ground & Power Bounce

The low impedance epitaxial p+ layer of the process essentially shorts out all of the chip grounds, creating a single ground network. Bypass capacitance can minimize bounce on the power lines, but ground is the global chip reference voltage and must be stabilized. To minimize absolute bounce, the ground network should contain as many pins as possible. A total of 28 pins were allocated, or 21% of all the package's pins. Simulations show that with the maximum number of I/O's switching, the ground bounce is between -540mV and +410mV at the maximum V_{DDIO} of 4V, which provides sufficient margin to prevent the ESD diodes from turning on (0.6-0.7V).

While bypass capacitance minimizes the V_{DDIO} bounce relative to ground to minimize I/O delay variation, it is also important to minimize the V_{DDIO} bounce in absolute terms to prevent the ESD diodes from turning on. Simulations demonstrated a worst case bounce of -600mV and +450mV at 4V for ten V_{DDIO} pins, with 2nF of on-chip bypass capacitance. The primary cause for the drop is the speed at which the output drivers are turned on. The device sizes were reduced to slow down the rise/fall times by 4x, but have fast turn off times.

The processor core is much more sensitive to power bounce due to timing considerations. A global reduction in V_{DD} will not affect functionality, as all the processor circuits' delay will scale appropriately. However, if only a part of the chip experiences a reduction in V_{DD}, timing violations leading to functional failure may occur. Thus, 16 pins were allocated to V_{DD}, and evenly distributed around the chip periphery. In addition, 16nF of bypass capacitance on V_{DD} is spread throughout the chip to minimize localized V_{DD} variations.

Global Routing

The RC delay on global signals is only critical on the processor bus, which has as little as 8ns to operate within at 4V. The RC delay product goes up with the square of the wire length, and becomes significant above 3.5mm for *Metal2*, and 7mm for *Metal3*. To provide sufficient margin at the maximum voltage of 4V, the RC delay must be kept below 500ps for all signal routes. The resistance is lowest on *Metal3*, which must be used for all long routes, due to its 50% lower RC delay. The program routeCap was written to calculate the RC for varying widths given a constant pitch, and report the width and space required for the wire to meet the maximum RC delay constraints. The longest route at 12mm required twice minimum width and spacing.

To eliminate Miller capacitance from adjacent parallel lines, the input and output busses are interleaved. Since they do not transition at the same time, any wire's nearest neighbors will not be switching concurrently, thereby negating the Miller effect.

6.3 Memory IC

The memory chip is based upon the SRAM design used within the processor's cache, and was designed to be DVS compatible while optimizing energy-efficiency. The new, key design challenge was the organization of the block-level architecture and global routing in order to minimize energy consumption. In addition, the memory chip supports split internal/external voltage sources, so that the I/O can be operated at a fixed voltage while the internal voltage varies in order to facilitate system debuggingThe SRAM die, shown in Figure 6.26, measures 9.6 x 10.4mm and contains 3.4M transistors.

6.3.1 Architecture

The total SRAM chip size of 64kB was set strictly by die size limitations. The basic SRAM block size was set to 1kB to provide a balanced trade-off of area efficiency (78% utilization) and energy consumption (50 pF/access). A flat hierarchy would be prohibitively expensive due to the large amount of capacitance on the bitlines, and the enormous drivers required within each SRAM block to drive this bus. Thus, a two-level hierarchy was chosen, in which the blocks are organized into an 8kB module, which was then replicated eight times for a total of 64kB, as shown in Figure 6.27.

The controller is responsible for interfacing with the processor bus, and contains additional circuitry to increment the internal address for burst-mode accesses, and to

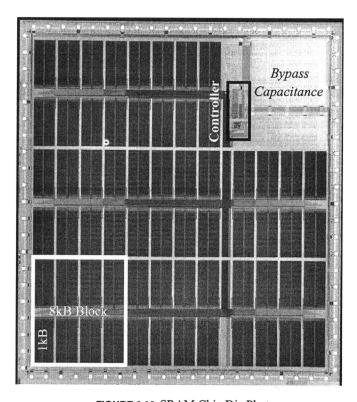

FIGURE 6.26 SRAM Chip Die Photo.

allow read-modify-write operations for byte writes. Also, the controller provides the SRAM control signals which get routed to each module. Since the address and data arrive on the same bus, only the lower 16 bits which contain the address information are routed to the address incrementer block.

Operation

The internal timing of the SRAM chip is shown in Figure 6.28. The address is not available on the internal address bus, *intAdd*, until almost the end of the first cycle, which necessitates the SRAM chip to always assert the *PWait* signal for one cycle until the first data word has been read. Subsequent reads can be performed without the need for asserting *PWait*, such that an eight-word cache line requires only ten

Memory IC

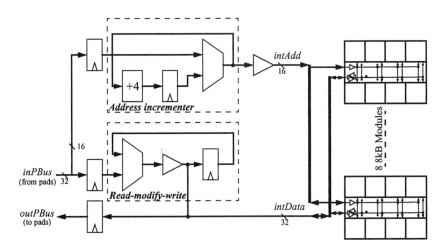

FIGURE 6.27 SRAM Architecture.

cycles to transfer across the processor bus. There is no need to stall the processor bus during either a word write, or a byte write, as shown in the lower two timing diagrams.

SRAM Module

The basic 1kB SRAM block in the module is essentially the same which was used in the processor cache and is replicated in the SRAM chip, with the addition of an address decoder. Since the 8kB module size is the same as the cache partition of 8kB, the output drivers see the same load and did not require redesign. The capacitance/cycle of the 1kB SRAM block is 50 pF/cycle for both read and write operations, with another 10 pF/cycle required to drive the interconnect within the module.

To reduce loading on the global busses and control signals, they are buffered before driving the local module interconnect. Since the data bus is bidirectional, it contains bidirectional transceivers which switch direction depending on whether it is a read or write. A potential hazard arises for read-modify-write operations required for byte writes, which switch the direction of the transceivers between cycles. To eliminate unnecessary short-circuit current, the enable signals have fast de-assertion times, and slow assertion times to ensure that either the local module data bus, or the global bus, *intData*, are not driven by two different transceivers at the same time. Furthermore, to reduce unnecessary switching activity, the direction of the transceivers are left in their last state, so that there is no default direction to return to.

Microprocessor and Memory IC's

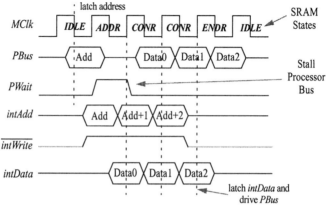

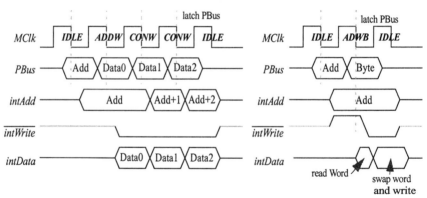

FIGURE 6.28 Internal SRAM Timing.

6.3.2 Energy Consumption

Performing a single read or write has an effective switched capacitance of 150-200pF over three and two bus clock cycles, respectively. Additional words in a burst read or write contribute approximately 75pF per word, and are much less because the internal address and control lines remain driven from the first data access. A byte write operation has an effective switched capacitance of 250pF, due to the combination of an SRAM read and write to complete it. The most common type of operation is a cache-line reload, which requires 725pF over ten bus cycles. Thus, if the SRAM chip is constantly active, it contributes a maximum of 36 pF/processor-cycle (there are at least two processor cycles per bus cycle), which is only 11% of the 320 pF/cycle consumed by the processor chip while it is active. In practice, the average capacitance/cycle will be lower since the SRAM is not constantly active. Thus, the SRAM was successfully designed to have minimal impact on total system energy consumption.

References

[6.1] Advanced RISC Machines, Ltd., *ARM Architecture and Implementation Reference*, Document Number ARM-DDI-0100A-I, Feb. 1996.

[6.2] Advanced RISC Machines, Ltd., *ARM 8 Data Sheet*, Document Number ARM-DDI-0100A-I, Feb. 1996.

[6.3] J. Hennessy, D. Patterson, *Computer Architecture: A Quantitative Approach*, Morgan Kaufmann, San Francisco, 1995.

[6.4] T. Burd, B. Peters, *A Power Analysis of a Microprocessor: A Study of an Implementation of the MIPS 3000 Architecture*, ERL Technical Report, University of California, Berkeley, 1994.

[6.5] A. Burstein, *Speech Recognition for Portable Multimedia Terminals*, Ph.D. Thesis, University of California, Berkeley, Document No. UCB/ERL M97/14, 1997.

[6.6] T. Pering, *Energy-Efficient Operating System Techniques*, Ph.D. Thesis, University of California, Berkeley, 2000.

[6.7] Hewlett Packard, *CMOS 14TA/B Reference Manual*, Document Number #A-5960-7127-3, Jan. 1995.

CHAPTER 7
DC-DC Voltage Conversion

As seen in previous chapters, use of the appropriate voltage for the performance required is one of the most powerful techniques for improving energy efficiency. This can be either done statically by choosing the single most appropriate voltage for the throughput required or dynamically by using information obtained by the operating system and applications. This chapter introduces switching regulators which allow translations of DC voltages with efficiencies of over 90%. Design equations and closed-form expressions for losses are presented for the three basic low-voltage CMOS switching regulator topologies – buck, boost, and buck-boost – controlled via pulse-width or pulse-frequency modulation. This then will be followed for the enhancements to the basic design that are required for efficient dynamic voltage regulation.

7.1 Introduction to Switching Regulators

The switching regulator shown in Figure 7.1 converts an unregulated battery source voltage V_{in} to the desired regulated DC output voltage V_o. A single-throw, double-pole switch chops V_{in} producing a rectangular wave having an average voltage equal to the desired output voltage. A low-pass filter passes this DC voltage to the output while attenuating the AC ripple to an acceptable value. The output is regulated by

DC-DC Voltage Conversion

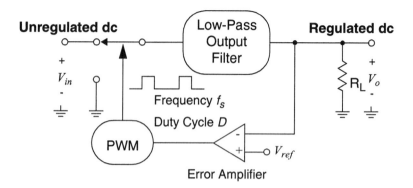

FIGURE 7.1 Block diagram of a PWM switching DC-DC converter.

comparing V_o to a reference voltage, V_{ref}, and adjusting the fraction of the cycle for which the switch is shorted to V_{in}. This pulse-width modulation (PWM) controls the average value of the chopped waveform, and thus controls the output voltage. A switching regulator has an efficiency which approach 100% as the components are made more ideal, however in practice, efficiencies above 75% are typical, and efficiencies above 90% are attainable.

There are several simple alternative arrangements of the switching and filter components that can be used to produce an output voltage larger or smaller than the input voltage, with the same or opposite polarity. Some of these will be discussed below. However, many of the design issues are similar, so first one topology, the step-down (buck) converter, will be discussed in more detail.

7.1.1 Buck Converter

The power train of the low-output-voltage buck circuit, which can produce any arbitrary output voltage $0 \leq V_o \leq V_{in}$, is given in Figure 7.2. The basic PWM operation is as follows: The power transistors (pass device M_p and rectifier M_n) chop the battery input voltage V_{in} to reduce the average voltage. This produces a square wave of variable duty cycle D and constant period $T_s = f_s^{-1}$ at the inverter output node, v_x. A typical periodic steady-state $v_x(t)$ waveform is shown in Figure 7.3. The second-order low-pass filter (L_f and C_f) passes the desired DC component of this chopped signal, while attenuating the AC to an acceptable ripple value. In the ideal case, the DC output voltage is given by the product of the input voltage and the duty cycle:

PWM Operation

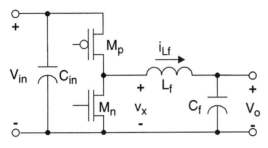

FIGURE 7.2 Low-output-voltage buck circuit

$$V_o = V_{in} \cdot D \qquad \text{(EQ 7.1)}$$

The switching pattern of M_n and M_p is pulse-width modulated, adjusting the duty cycle of the rectangular wave at v_x, and ultimately, the DC output voltage, to compensate for input and load variations. The pulse-width modulation is controlled by a negative feedback loop, shown in the block diagram of Figure 7.1, but omitted from Figure 7.2 for simplicity.

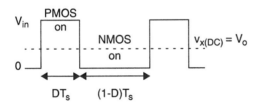

FIGURE 7.3 Nominal periodic steady-state $v_x(t)$ buck circuit waveform

7.2 PWM Operation

Figure 7.4 shows the steady-state operating waveforms of the buck circuit in PWM operation. The switching cycle is initiated when PMOS device, M_p, turns on. During the interval, D, of the switching period, T_s, the inverter output node, v_x, is shorted to V_{in}. A constant positive potential, V_{in}-V_o, is applied across the inductor, and i_{Lf} lin-

DC-DC Voltage Conversion

early increases from its minimum value to its maximum value. Some of the energy removed from the battery is stored in the magnetic field of the inductor, and some is delivered to the filter capacitor and the load.

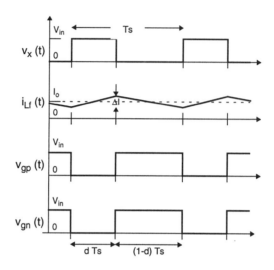

FIGURE 7.4 Periodic steady-state PWM waveforms for the buck circuit.

Then, the PMOS device is turned off, and the NMOS rectifier device, M_n, is turned on to pick up the inductor current, shorting v_x to ground. During this interval, (1-D) of the cycle, a constant negative potential is applied across the inductor, and i_{Lf} linearly decreases from its maximum value to its minimum value. Excess energy in the inductor is delivered to the output filter capacitor and load. The cycle then repeats by turning off M_n and turning on M_p.

In periodic steady-state, regulation is maintained when the charge drawn from the battery during a switching period is equal to the charge consumed by the load.

7.2.1 Output Filter Design

In Figure 7.5, the rectangular wave of the inverter output node is applied to the second order low-pass output filter of the buck circuit (L_f and C_f) which passes the

PWM Operation

desired DC component of v_x while attenuating the AC component to an acceptable ripple value. Load R_L draws a DC current I_o from the output of the filter. Figure 7.6 shows the nominal steady-state $i_{Lf}(t)$ and $v_o(t)$ waveforms for a rectangular input $v_x(t)$.

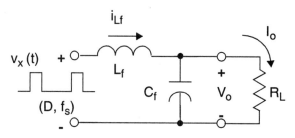

FIGURE 7.5 The output filter of the buck circuit (L_f and C_f) with load R_L.

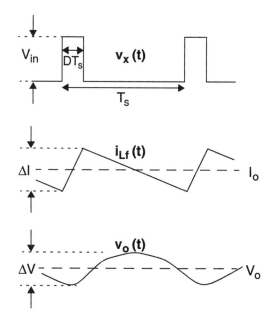

FIGURE 7.6 Nominal steady-state waveforms of the buck circuit output filter.

DC-DC Voltage Conversion

DC-DC Voltage Conversion

In order to achieve the large attenuation needed in a practical power circuit, $L_f \cdot C_f \gg \omega_s^{-2}$, where $\omega_s = 2\pi f_s$, and f_s is the switching frequency of the converter. In this case, the filter components may be sized independently, using time domain analysis, rather than frequency domain analysis. Neglecting the effects of output voltage ripple ($v_{o-AC} \ll v_{x-AC}$, for a rectangular input with period T_s, the AC inductor current waveform is triangular with period T_s and peak-to-peak ripple ΔI symmetric about the average load current I_o. The peak-to-peak current ripple may be found by integrating the AC component of the $v_x(t)$ waveform over a fraction, D, of one cycle, yielding:

$$\Delta I = \frac{V_{in} \cdot D \cdot (1-D)}{L_f \cdot f_s} = \frac{V_o \cdot (1-D)}{L_f \cdot f_s} \qquad \text{(EQ 7.2)}$$

The output filter capacitor is selected to ensure that its impedance at the switching frequency, including its equivalent series resistance (ESR), is small relative to the load impedance. Thus, the AC component of the inductor current flows into the filter capacitor, rather than the load. For many capacitor technologies at frequencies above several hundred kilohertz, the resistive impedance dominates over the capacitive impedance. In high-current-ripple designs, a primary design goal is to minimize ESR to reduce both output voltage ripple and conduction loss (see below). For this reason, a high-Q capacitor technology, such as multilayer ceramic, is typically used, and even at high frequencies, ESR may be neglected in calculating output voltage ripple. Considering only capacitive impedance, the peak-to-peak output voltage ripple may be found through charge conservation. Assuming the AC inductor current flows only into the filter capacitor:

$$\Delta V = \frac{\Delta I}{8 \cdot C_f \cdot f_s} = \frac{V_o \cdot (1-D)}{8 \cdot L_f \cdot C_f \cdot f_s^2} \qquad \text{(EQ 7.3)}$$

This output voltage ripple is symmetric about the desired DC output voltage V_o, and, for the $v_x(t)$ waveform shown in Figure 7.6, is piecewise quadratic with period T_s.

Equation 7.2 and Equation 7.3 illustrate the two principle means of miniaturizing a DC-DC converter. First, it can be readily seen that the necessary values of filter inductance and capacitance decrease with f_s^{-1}. Thus, a higher operating frequency typically results in a smaller converter. Second, because the requirement of interest is output voltage ripple, it is the $L_f \cdot C_f$ product, rather than the values of the individual components, that is important. Through choice of a higher current ripple, ΔI, a lower filter inductance solution may be obtained, often resulting in a smaller regulator.

7.2.2 Sources of Dissipation

The power train of the low-output-voltage buck circuit, including all series resistance, parasitic capacitance C_x, stray inductance L_s, and drain-body diodes of the power transistors, is shown in Figure 7.7. Listed below are the chief sources of dissipation that cause the conversion efficiency of this circuit to be less than unity.

Conduction Loss

Current flow through non-ideal power transistors, filter elements, and interconnections results in dissipation in each component:

$$P_q = i_{rms}^2 \cdot R \qquad \text{(EQ 7.4)}$$

where i_{rms} is the root mean squared current through the component, and R is the resistance of the component.

In PWM mode, the rms current has a DC and an AC component:

$$i_{rms}^2 = i_{rms(DC)}^2 + i_{rms(AC)}^2 \qquad \text{(EQ 7.5)}$$

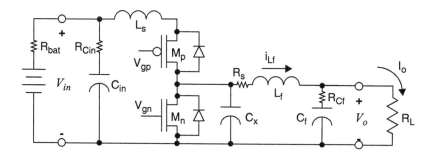

FIGURE 7.7 Low-output-voltage buck circuit, including parasitics.

DC-DC Voltage Conversion

where:

$$i^2_{rms(DC)} = d \cdot I_o^2 \quad \text{(EQ 7.6)}$$

and

$$i^2_{rms(AC)} = d \cdot \frac{1}{3} \cdot \left(\frac{\Delta I}{2}\right)^2 \quad \text{(EQ 7.7)}$$

Here, $0 \le d \le 1$ is a weighting factor which indicates the duty cycle of current flow through the component, I_o is the DC load current, and ΔI is the peak-to-peak inductor current ripple.

While DC conduction loss scales quadratically with decreasing load current, AC conduction loss is a fixed quantity and may substantially degrade efficiency at light load.

Gate-Drive Loss

Raising and lowering the gate of a power transistor each cycle dissipates an average power:

$$P_g = E_g \cdot f_s \quad \text{(EQ 7.8)}$$

where E_g is directly proportional to the gate energy transferred per off-to-on-to-off gate transition cycle (which can include some energy due to Miller effect), and includes dissipation in the drive circuitry.

Gate-drive loss is independent of load current and will therefore degrade light-load efficiency.

Timing Errors

Three mutually exclusive mechanisms of loss attributed to timing errors in the switching of the power MOSFETs are described below. Each is independent of load.

i) No Dead-Time: Short Circuit Loss:

A short-circuit path may exist temporarily between the input rails during power FET switching transitions. To avoid potentially large short-circuit losses, it is necessary to provide dead-times in the conduction of the MOSFETs to ensure that the two devices never conduct simultaneously.

ii) Dead-Times Too Long: Body-Diode Conduction:

If the durations of the dead-times are too long, the body diode of the NMOS power transistor may be forced to pick up the inductor current for a fraction of each cycle. Since in low-voltage applications, the forward bias diode voltage ($V_d \approx 0.7$ V) can be comparable to the output voltage, its conduction loss may be significant:

$$P_{diode} \approx 2 \cdot I_o \cdot V_d \cdot t_{err} \cdot f_s \qquad \text{(EQ 7.9)}$$

where t_{err} is the timing error between complementary power MOSFET conduction intervals.

Furthermore, when the PMOS device is turned on, it must remove the excess minority carrier charge from the body diode, dissipating an energy bounded by:

$$E_{rr} = Q_{rr} \cdot V_{in} \qquad \text{(EQ 7.10)}$$

where Q_{rr} is the stored charge in the body diode.

iii) Dead-Times Too Short: Capacitive Switching Loss:

In a hard-switched converter, MOSFET M_p charges parasitic capacitance C_x to V_{in} each cycle, dissipating an average power:

$$P_{Cx(LH)} = \frac{1}{2} \cdot C_x \cdot V_{in}^2 \cdot f_s \qquad \text{(EQ 7.11)}$$

where C_x includes reverse-biased drain-body junction diffusion capacitance C_{db} and some or all of the gate-drain overlap (Miller) capacitance C_{gd} of the power transistors, wiring capacitance from their interconnection, and stray capacitance associated with L_f. In ultra-low-power monolithic converters, C_x may be dominated by parasitics associated with the connection of an off-chip filter inductor, which include a bond pad, bond wire, pin, and board interconnect capacitance.

When M_p is turned off, the inductor begins to discharge C_x from V_{in} to ground. If M_n is turned on exactly when v_x reaches ground, this transition is lossless. If the NMOS device is turned on too late, v_x will be discharged below ground, until the body diode is forced to conduct (see above). If the NMOS device is turned on too early, it will discharge v_x to ground through its channel, introducing losses:

$$P_{Cx(HL)} = \frac{1}{2} \cdot C_x \cdot v_x^2 \cdot f_s \leq \frac{1}{2} \cdot C_x \cdot V_{in}^2 \cdot f_s \quad \text{(EQ 7.12)}$$

Stray Inductive Switching Loss

Energy storage by the stray inductance L_s in the loop formed by the input decoupling capacitor C_{in} and the power transistors causes dissipation (Figure 7.8). Here, M_p and M_n are modeled as ideal switches, and L_f is modeled as a current source of value $i(t) = i_{Lf}(t)$. When switch M_p closes, it charges L_s from $i_{Ls} = 0$ to $i_{Ls} = I_{min}$. When M_p opens and M_n closes, L_s is discharged from $i_{Ls} = I_{max}$ to $i_{Ls} = 0$. The average power dissipation is equal to:

$$P_{Ls} = \frac{1}{2} \cdot L_s \cdot (I_{min}^2 + I_{max}^2) \quad \text{(EQ 7.13)}$$

This loss is somewhat dependent on load current, as:

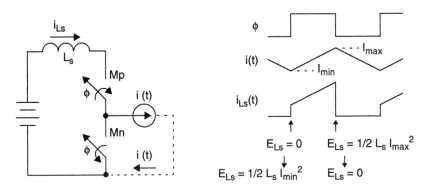

FIGURE 7.8 Energy dissipation due to stray inductance.

$$I_{min} = I_o - \frac{\Delta I}{2} \qquad \text{(EQ 7.14)}$$

and

$$I_{max} = I_o + \frac{\Delta I}{2} \qquad \text{(EQ 7.15)}$$

The value of L_s is dependent on PCB layout, packaging, bonding, and chip layout, and is reduced by minimizing the area of this critical high current loop. In a multi-layer interconnection technology, the lowest stray inductance is achieved by using a conductor that overlaps a return path in a different layer, with thin dielectric separating the layers. In a careful design:

$$1 \text{ nH} < L_s < 10 \text{ nH} \qquad \text{(EQ 7.16)}$$

Quiescent Operating Power

The PWM and other control circuitry consume static power. In low-power applications, this control power may contribute substantially to the total losses, even at full-load.

7.3 PFM Operation

While a PWM DC-DC converter can be made to be highly efficient at full load, many of its losses are independent of load current, and it may, therefore, dissipate a significant amount of power relative to the output power at light loads. Figure 7.9 plots total losses versus a 1000:1 load range for a typical PWM buck converter. As the load scales downward, AC conduction loss, switching loss, and PWM control power become increasingly significant, and total dissipation in the converter asymptotes to a fixed minimum power dissipation. From this plot, it may be concluded that a PWM converter which is 94% efficient at full load is roughly 3% efficient at one thousandth full load. If the converter is used at full load for little of its operating time, energy loss at light load will be the dominant limitation on battery run-time, and improving efficiency at light load becomes essential.

One control scheme which achieves high efficiency over a wide load range is pulse-frequency modulation (PFM). In this scheme, conceptually illustrated in Figure 7.10,

DC-DC Voltage Conversion

FIGURE 7.10 A conceptual illustration of PFM control.

the converter is operated only in short bursts at light load. Between bursts, both power FET's are turned off, and the circuit idles with zero inductor current. During this period, the output filter capacitor sources the load current. When the output is dis-

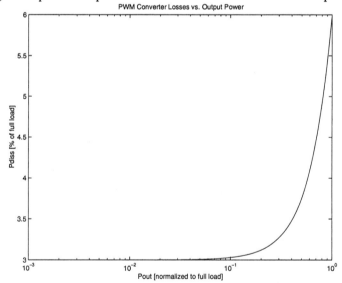

FIGURE 7.9 PWM converter losses vs. load.

charged to a certain threshold below V_{REF}, the converter is activated for another burst, returning charge to C_f. Thus, the load-independent losses in the circuit are reduced. As the load current decreases, the idle time increases. Regulation is maintained when the charge delivered through the inductor is equal to the charge consumed by the load.

7.3.1 Output Filter Design

Figure 7.11 shows the steady-state buck circuit waveforms under PFM control. The

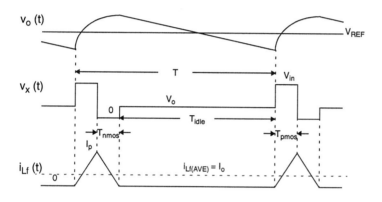

FIGURE 7.11 Steady-state PFM waveforms.

PFM operation is described heuristically in Figure 7.10: When the output voltage drops to a certain threshold below V_{REF} (likely sensed by a hysteretic comparator), a burst of charge is delivered, returning V_o to a threshold above V_{REF}. (Unlike the waveforms of Figure 7.10, here, only a single switching event of the DC-DC converter is used to deliver each burst of charge.) This charge burst is delivered with high energy efficiency through the inductor as follows: The PMOS device is turned on for a time interval, T_{pmos}. Some of the energy removed from the battery is delivered to the output; the rest is stored in the inductor. During this interval, the inductor current slews at a rate of:

$$\frac{di_L}{dt} = \frac{(V_{in} - V_o)}{L_f} \qquad \text{(EQ 7.17)}$$

DC-DC Voltage Conversion

and reaches its peak value of $i_{Lf} = I_p$ at the conclusion of the PMOS conduction interval. The PMOS device is then turned off, and after a short dead-time, the NMOS device is turned on to pick up the inductor current. During NMOS conduction, v_x is shorted to ground, and the energy stored in the inductor is released to the output. The inductor current slews from I_p to 0 at a rate of:

$$\frac{di_L}{dt} = \frac{-V_o}{L_f} \qquad \text{(EQ 7.18)}$$

The NMOS device is (ideally) turned off when i_{Lf} decays to zero. At this time, v_x will ring up to V_o, and the circuit will idle with zero inductor current and the output capacitor sourcing the load current.

The total charge delivered through the inductor by each PFM burst is found by integrating the area under the $i_{Lf}(t)$ waveform for one switching cycle of the DC-DC converter:

$$Q_L = \frac{1}{2} \cdot I_p \cdot (T_{pmos} + T_{nmos}) \qquad \text{(EQ 7.19)}$$

Because a time delay is fairly straightforward to implement on-chip, a convenient PFM controlling variable is the PMOS conduction interval, T_{pmos}. The NMOS conduction interval is uncontrolled, but can be found in relation to the controlling variable by equating the products of the linear inductor current slopes and the conduction intervals to the peak current, I_p:

$$I_p = \frac{(V_{in} - V_o) \cdot T_{pmos}}{L_f} = \frac{V_o \cdot T_{nmos}}{L_f} \qquad \text{(EQ 7.20)}$$

$$T_{nmos} = \frac{(V_{in} - V_o)}{V_o} \cdot T_{pmos} \qquad \text{(EQ 7.21)}$$

In terms of only the controlled variable,

$$Q_L = \frac{1}{2} \cdot \frac{T_{pmos}^2 \cdot (V_{in} - V_o) \cdot V_{in}}{V_o \cdot L_f} \qquad \text{(EQ 7.22)}$$

Regulation is maintained when this delivered charge is equal to the charge consumed by the load:

$$Q_L = I_o \cdot T \qquad \text{(EQ 7.23)}$$

where

$$T = T_{idle} + T_{pmos} + T_{nmos} \qquad \text{(EQ 7.24)}$$

is the variable PFM repetition period.

Inductor Value

To support a maximum load current, $I_{o(max)}$:

$$L_f = \frac{T_{pmos} \cdot (V_{in} - V_o)}{2 \cdot I_{o(max)}} \qquad \text{(EQ 7.25)}$$

As indicated by Equation 7.22, a smaller value of inductance than that given in Equation 7.25 will support a larger load current, and will support $I_{o(max)}$ with a larger time between pulses, T_{idle}.

Capacitor Value

The capacitor is selected to ensure that the peak-to-peak output voltage ripple, ΔV, is maintained to a certain percentage of V_o. The worst-case output voltage ripple is calculated assuming that all of the charge delivered through the inductor is absorbed by C_f:

$$C_f = \frac{Q_L}{\Delta V_f} \qquad \text{(EQ 7.26)}$$

7.3.2 Sources of Dissipation

The mechanisms of loss in PFM operation are identical to those presented in Section 7.2.2 for PWM operation. However, PFM converters are shut down during the idle time, T_{idle}, between pulses and, with the exception of some static dissipation in the control circuits, dissipate energy only during pulses. Thus, the analysis below presents losses in terms of the energy dissipated per PFM pulse.

DC-DC Voltage Conversion

Assuming a small AC voltage ripple $\Delta V \ll V_o$, the energy delivered to the load in one PFM pulse is given by:

$$E_{pulse} = Q_L \cdot V_o \qquad \text{(EQ 7.27)}$$

The overall efficiency of the converter in PFM operation is then expressed as the ratio given by:

$$\eta = \frac{E_{pulse}}{E_{pulse} + E_{diss}} \qquad \text{(EQ 7.28)}$$

Conduction Loss

Current flow through non-ideal power transistors, filter elements, and interconnections results in energy dissipation in each component:

$$E_q = \int_0^{T_{pulse}} i(t)^2 R \, dt \qquad \text{(EQ 7.29)}$$

where $i(t)$ is the current through the component, $T_{pulse} = T_{pmos} + T_{nmos}$, and R is the resistance of the component.

Gate-Drive Loss

Raising and lowering the gate of a power transistor each cycle dissipates an energy E_g. This is directly proportional to the gate energy transferred per off-to-on-to-off gate transition cycle (which can include some energy due to Miller effect), and includes dissipation in the drive circuitry.

Switch Transitions and Timing Errors

PMOS Turn-On:

The power PMOS device is always turned on with the converter idling – in steady-state, $v_x = V_o$ and $i_{Lf} = 0$. The energy stored on C_x just prior to PMOS turn-on is:

$$E_{Cx(initial)} = \frac{1}{2} \cdot C_x \cdot V_o^2 \qquad \text{(EQ 7.30)}$$

PFM Operation

The PFM switching cycle is initiated when M_p charges C_x from $v_x = V_o$ to $v_x = V_{in}$. The energy stored on C_x just after this transition is:

$$E_{Cx(final)} = \frac{1}{2} \cdot C_x \cdot V_{in}^2 \qquad \text{(EQ 7.31)}$$

The energy drawn from the battery during this transition is equal to:

$$E_{in} = V_{in} \cdot \Delta Q_{Cx} = V_{in} \cdot C_x \cdot (V_{in} - V_o) \qquad \text{(EQ 7.32)}$$

The energy dissipated in the turn-on transition is therefore given by:

$$E_{Cx(IH)} = E_{in} - (E_{Cx(final)} - E_{Cx(initial)}) = \frac{1}{2} \cdot C_x \cdot (V_{in} - V_o)^2 \qquad \text{(EQ 7.33)}$$

where the IH subscript denotes the idle-to-high transition at v_x.

PMOS Turn-Off, NMOS Turn-On:

The PMOS off to NMOS on transition is nearly identical to that in PWM mode (Section). With no dead-time provided, a short-circuit path may exist temporarily during switch transitions, introducing significant loss. If the dead-time is too short, M_n discharges C_x through its resistive channel, introducing a loss bounded by:

$$E_{Cx(HL)} \leq \frac{1}{2} \cdot C_x \cdot V_{in}^2 \qquad \text{(EQ 7.34)}$$

(The subscript HL indicates the high-to-low transition at v_x.)

If the dead-time is too long, the inductor discharges C_x below ground, until the NMOS body diode becomes forward-biased.

NMOS Turn-Off:

Ideally, the NMOS device is gated off when i_{Lf} decays to zero. In this case, the $i_{Lf}(t)$ and $v_x(t)$ waveforms will ring from the initial condition, $i_{Lf}(t) = 0$, $v_x(t) = 0$, to the final steady-state condition during idle mode, $i_{Lf}(t) = 0$, $v_x(t) = V_o$ in the resonant circuit of Figure 7.12. Since in any practical DC-DC converter, $C_f \gg C_x$, in this circuit the output capacitor is modeled as an ideal voltage source. The ringing $v_x(t)$ and $i_{Lf}(t)$ waveforms are shown in Figure 7.13.

The energy dissipated in this ring (in the equivalent series resistance in the L_f-C_x-C_f tank, R) is fundamentally equal to:

$$E_{Cx(LI)} = \frac{1}{2} \cdot C_x \cdot V_o^2 \qquad \text{(EQ 7.35)}$$

The LI subscript in Equation 7.35 indicates the low-to-idle transition at v_x. Note that if:

$$v_{x(max)} = 2V_o > V_{bat} + V_D \qquad \text{(EQ 7.36)}$$

where V_D is the PMOS forward bias diode voltage (approximately equal to 0.7 V), the PMOS body diode will conduct for a portion of the first sinusoidal cycle, dissipating additional energy.

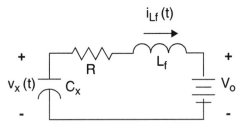

FIGURE 7.12 Resonant tank during PFM idle time interval.

If the NMOS device turns off too early ($i_{Lf} = I_\varepsilon > 0$), additional energy stored in the output inductor is dissipated. For:

$$E_L = \frac{1}{2} \cdot L \cdot I_\varepsilon^2 < E_C = \frac{1}{2} \cdot C_x \cdot V_D^2 \qquad \text{(EQ 7.37)}$$

where V_D is the forward bias NMOS diode voltage (also approximately equal to 0.7 V), the NMOS body diode will not forward bias, and all of E_L will be dissipated in the resistance in series with the LC tank. If the condition of Equation 7.37 is not satisfied, the NMOS body diode will conduct, dissipating some of E_L and delivering the rest to the output. Figure 7.14 and Figure 7.15 show the equivalent circuit and $i_{Lf}(t)$ and $v_x(t)$

PFM Operation

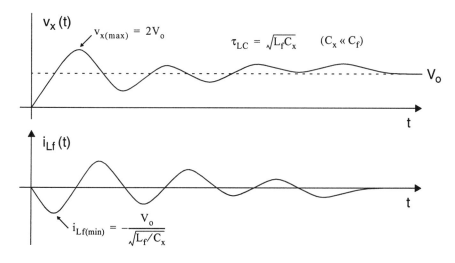

FIGURE 7.13 LC ring after NMOS turn-off.

waveforms during NMOS body diode conduction. Since the voltage drop across the

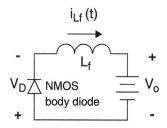

FIGURE 7.14 Equivalent circuit during NMOS body diode conduction.

diode is large compared to that across any resistance in series with the LC tank, R is eliminated from this model, leaving the body diode as the only dissipater. In this case, the ratio of energy dissipated to energy stored is equal to the ratio of voltage drop across the diode to that across the inductor:

$$E_{diode} = E_L \cdot \frac{V_D}{V_{bat} + V_D - V_o} \qquad \text{(EQ 7.38)}$$

DC-DC Voltage Conversion

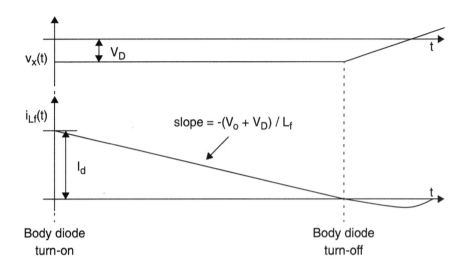

FIGURE 7.15 Waveforms during NMOS body diode conduction.

where

$$E_L = \frac{1}{2} \cdot L \cdot I_d^2 \qquad \text{(EQ 7.39)}$$

and

$$I_d = I_\varepsilon - \frac{V_D}{\sqrt{L_f/C_x}} \qquad \text{(EQ 7.40)}$$

In addition,

$$E_R = \frac{1}{2} \cdot C_x \cdot (V_o^2 + V_D^2) \qquad \text{(EQ 7.41)}$$

is dissipated in the series R before and after body diode conduction, resulting in an

PFM Operation

energy penalty of $\frac{1}{2} \cdot C_x \cdot V_D^2$.

From the above results, the total energy penalty associated with an early NMOS turn-off transition is:

$$E_{penalty} = \frac{1}{2} \cdot L \cdot I_\varepsilon^2 \quad \text{for} \quad I_\varepsilon < \frac{V_D}{\sqrt{L_f/C_x}} \quad \text{(EQ 7.42)}$$

otherwise

$$E_{penalty} = \frac{1}{2} \cdot C_x \cdot V_D^2 + \frac{1}{2} \cdot L \cdot I_d^2 \cdot \frac{V_D}{V_o + V_D}$$

If the NMOS device turns off too late ($i_{Lf} = I_\varepsilon < 0$) some or all of the energy stored in the inductor is dissipated in the series resistance and/or the PMOS body diode. Since the analysis is similar to the derivation of Equation 7.42, only the resulting losses are given:

$$E_{penalty} = \frac{1}{2} \cdot L \cdot I_\varepsilon^2 \quad \text{for} \quad |I_\varepsilon| < \frac{(V_D + V_{bat})}{\sqrt{L_f/C_x}} \quad \text{(EQ 7.43)}$$

otherwise

$$E_{penalty} = \frac{1}{2} \cdot C_x \cdot (V_D + V_{bat})^2 + \frac{1}{2} \cdot L \cdot I_d^2 \cdot \frac{V_D}{V_{bat} + V_D - V_o}$$

In Equation 7.43,

$$I_d = I_\varepsilon + \frac{(V_D + V_{bat})}{\sqrt{L_f/C_x}} \quad \text{(EQ 7.44)}$$

and is less than zero.

Stray Inductive Switching Loss

Energy storage by the stray inductance L_s in the loop formed by the input decoupling capacitor C_{in} and the power transistors causes dissipation (Figure 7.8). In the PFM PMOS turn-on transition, $i_{Lf} = 0$, and since no energy is stored in L_s, there is no associated loss. The PMOS turn-off / NMOS turn-on transition occurs when the peak

DC-DC Voltage Conversion

DC-DC Voltage Conversion

inductor current, I_p, flowing into the power circuit is switched from the high-side to the low-side input terminal, introducing a loss equal to:

$$E_{Ls} = \frac{1}{2} \cdot L_s \cdot I_p^2 \qquad \text{(EQ 7.45)}$$

Quiescent Operating Power

The PFM control circuitry consumes static power, even when the converter is idling. The energy dissipation per charge burst is given by:

$$E_{static} = P_{static} \cdot T \qquad \text{(EQ 7.46)}$$

where T is the variable PFM repetition period.

This proves to be the fundamental limitation to light-load efficiency under PFM control. Since T increases with decreasing load, E_{static} becomes the dominant source of light-load loss. Effort must therefore be concentrated on minimizing this static power dissipation.

7.4 Other Topologies

Two other basic configurations for PWM switching converters are the boost converter (Figure 7.16 and the buck-boost converter (Figure 7.18). All three basic topologies – buck, boost, and buck-boost – are similar in that they each have two complementary switches and one inductor. Their conversion ratios may all be adjusted by varying the duty cycle with frequency held constant. They can all be derived from the same basic switching cell *[7.1]*.

The boost converter produces output voltages $V_o \geq V_{in}$. A typical steady-state $v_x(t)$ waveform is shown in Figure 7.17. In one portion of the cycle, (1-D), the NMOS device is on, and the input voltage is applied across L_f, building up current and thus storing energy in the inductor. When the NMOS switch is turned off, the attempt to interrupt the current in the inductor causes the voltage at node v_x to rise rapidly. The PMOS device is turned on at this point, limiting the voltage produced by this inductive kick to the voltage on the output capacitor. (If the PMOS device were not turned on, its drain-body diode would short v_x to one diode drop above V_o.) During the fraction of the cycle, D, that the PMOS device conducts, some of the energy stored in the

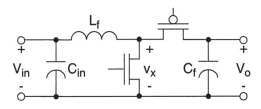

FIGURE 7.16 Low-voltage CMOS boost circuit.

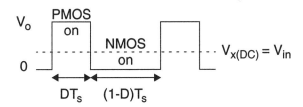

FIGURE 7.17 Nominal steady-state $v_x(t)$ boost circuit waveform.

inductor is transferred to the output, along with additional energy flowing from the input. The cycle then repeats.

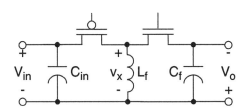

FIGURE 7.18 Low-voltage CMOS buck-boost circuit.

The boost converter may be considered a variation of the buck converter, but with power flow from the lower voltage side to the higher voltage side. The voltage at node

DC-DC Voltage Conversion

v_x is a rectangular wave whose DC component is equal to the input voltage. (It must be equal, as the average voltage across the inductor must be zero for periodic steady state.) Thus, the input and output voltages are related by:

$$V_{in} = V_o \cdot D \qquad \text{(EQ 7.47)}$$

the same relation as for the buck converter, but with the input and output terminals reversed.

The operation of the buck-boost converter (Figure 7.18) is similar to that of the buck converter, in that the cycle starts with the input voltage applied across the inductor, in this case through the PMOS device for a duration, $D \cdot T_s$. However, when the PMOS device is turned off, the voltage at v_x heads downward, and the circuit produces an output voltage polarity opposite to that of the input (Figure 7.19). The energy transferred to C_f during this portion, (1-D), of the cycle (while the NMOS device conducts) is only the energy stored in the inductor, with none coming directly from the input. Setting the average voltage across the inductor equal to zero allows the conversion ratio to be found:

$$V_o = V_{in} \cdot \frac{D}{1-D} \qquad \text{(EQ 7.48)}$$

Note that this allows input voltages of smaller or larger magnitude than the input, hence the name "buck-boost".

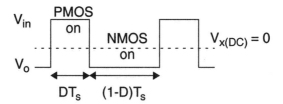

FIGURE 7.19 Nominal steady-state $v_x(t)$ buck-boost circuit waveform.

7.5 Dynamic Voltage Conversion

The basic architecture of a regulator which dynamically adjusts the voltage is a non-linear negative-feedback loop. which in steady-state operation forces the processor clock, to be:

$$f_{DES} = M \cdot (1 \text{ MHz}) \qquad \text{(EQ 7.49)}$$

where f_{DES} is the desired frequency and M is this value in units of 1 MHz which is the digital value stored by the processor hardware. Thus, the processor requires no knowledge of the actual supply voltage. It effectively adjusts V_{DD} by requesting a new operating frequency, f_{DES}.

The loop is built around a buck-converter which is amenable to high-efficiency, low-voltage regulation [7.2]. Using a digital PWM algorithm, the buck-converter converts the battery voltage, V_{BAT}, to the desired output voltage, V_{DD}, as a function of the pulse duty cycle, D:

$$V_{DD} = V_{BAT} \cdot D \qquad \text{(EQ 7.50)}$$

To improve the converter's efficiency at low voltage and/or light load, the converter loop also implements a PFM algorithm as described in Section 7.3, at low voltage and/or light load, the processor's energy consumption is very small, and so too is the current it draws from V_{DD}. Rather than enable the PMOS for an infinitesimally small amount of time to replace the small amount of charge removed from C_{DD} by the processor, the converter is selectively enabled through pulse-skipping, by which, in a given period T_S, if the voltage drop on V_{DD} is sufficiently small, the converter is simply disabled. The conversion efficiency is greatly increased due to the saved energy cost of enabling the power FETs, but comes with the penalty of increased voltage ripple.

7.5.1 Loop Architecture

The full converter loop architecture is shown in Figure 7.20. The output of the ring oscillator, f_{VCO}, clocks a counter which is reset at 1 MHz intervals. This provides the quantized digital word, f_{MEAS}, which is the measured clock frequency in MHz. This value is subtracted from the desired clock frequency, f_{DES}, to generate an error fre-

DC-DC Voltage Conversion

quency value, f_{ERR}. A positive value indicates a higher voltage is required to increase the processor clock, f_{CLK}, (a buffered version of f_{VCO}) and a negative value indicates that the voltage is too high.

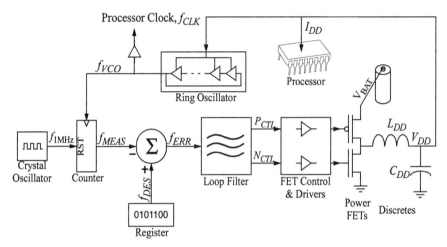

FIGURE 7.20 DVS Voltage Converter Loop Architecture.

The loop filter does two important functions. First, it converts the frequency error into an equivalent voltage error via a hardware look-up table. Next, it converts the equivalent voltage error into an update command for the power FETS through the hybrid PWM-PFM scheme described in the previous section. When $-3 \leq f_{ERR} \leq 0$ indicating that the voltage is slightly high, the pulse-skipping algorithm disables the converter for the current clock cycle, allowing the processor to discharge V_{DD}. Any other value of f_{ERR} enables the converter for the current cycle through the control signals P_{CTL} and N_{CTL} [7.3].

These two control signals are converted to power FET enable signals, which are buffered to drive the large gate capacitance of the power FETs. The buck converter produces an output voltage V_{DD} which is sent back to the ring oscillator, closing the loop. In addition, the processor is powered by V_{DD}, so it draws a time-varying current I_{DD} from the output capacitor.

7.5.2 Loop Stability

The external filter components, shown in Figure 7.21, primarily dictate the frequency response of the converter loop. R_{DD} is the effective resistance of the V_{DD} load, and varies as a function of V_{DD}.

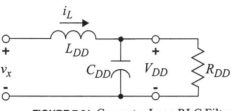

FIGURE 7.21 Converter Loop RLC Filter.

In a typical buck converter, this filter has two poles, due to the capacitor voltage, V_{DD}, and inductor current, i_L, state variables. However, in this system, charge is delivered to the capacitor, C_{DD}, in discrete quantities, thereby ensuring that i_L starts and ends each cycle at zero which eliminates it as a state variable. Although operating in this discontinuous mode increases the voltage ripple on C_{DD}, it reduces this filter to a one pole system, whose pole is set by R_{DD} and C_{DD}, as shown in Figure 7.22.

There is a sampling delay introduced by the front-end clock quantizer, which places another pole around 1 MHz. As V_{DD} increases, R_{DD} decreases, and the dominant pole moves higher in frequency, potentially resulting in instability. For the system implementation in Chapter 9, peak current was 125mA at 4V so that the dominant pole is a maximum of 7kHz and the loop gain is less than one at 1 MHz, thereby ensuring system stability.

7.5.3 Clock Generation

A significant benefit of the converter loop architecture in a DVS system is that it provides clock generation for the processor. The only external circuits required is a 1 MHz oscillator, which can be implemented with little power dissipation (<10μW [7.4]), with the power dissipated driving the clock signal on the printed-circuit board (PCB) given by:

DC-DC Voltage Conversion

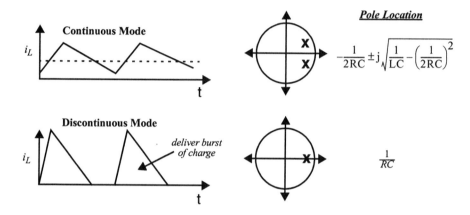

FIGURE 7.22 Reducing the Buck Converter to a One Pole System

$$P_{PCB} = f_{CLK} \cdot C_L \cdot V_{DD}^2 \qquad \text{(EQ 7.51)}$$

For a C_L of 10pF, and a V_{DD} of 3.3V, the power dissipation for driving this 1 MHz clock signal is 100μW.

A typical processor either generates the f_{CLK} on chip via a phase-locked loop (PLL), or uses an externally generated signal. For a 100 MHz clock signal, the power driving the 10pF would be 11mW at 3.3V, and the oscillator itself can add another 10-100mW. Most mid-to-high performance processors have an on-chip PLL to generate the processor clock signal. But even the lowest reported power dissipation is 1.5mW, and still requires an external 3.68 MHz crystal oscillator *[7.5]*.

In contrast, the ring-oscillator for the converter loop is the equivalent of 33 gates switching every cycle; in our 0.6μm process, this is approximately 1pF. The power dissipation of the ring oscillator scales with f_{CLK} and V_{DD}. At the low corner of 1.1V and 8 MHz, the power dissipated is only 10μW; this is 10x lower than conventional clock generation approaches, even taking the required external 1 MHz crystal oscillator into account. In addition, this capacitance will also scale down in technology so that in better process technology, the power dissipation will be lower for a given f_{CLK} and V_{DD}. Further reduction can be achieved by integrating the 1 MHz oscillator circuit on-chip, leaving only the crystal external to the chip, which would eliminate the power dissipation for driving the external 1 MHz clock signal.

7.5.4 Conversion Efficiency

The efficiency of a voltage regulator is defined as:

$$\eta = \frac{\text{Power Delivered to Load}}{\text{Total Power Dissipation}} \quad \text{(EQ 7.52)}$$

with 100% being the maximum efficiency possible, in which no power is lost in delivering energy to the load circuits. The buck converter is very efficient at voltage conversion, with efficiencies typically in the 90-95% range [7.2]. While it can be designed methodically for a fixed operating voltage, the difficulty arises in designing for this efficiency across a range of voltage and current loads. Several techniques have been developed for the converter loop design to improve the efficiency over this broad range of operating conditions [7.3].

The loop filter PWM-PFM algorithm will not deliver charge when $-3 \leq f_{ERR} \leq 0$. For low voltage and/or light load conditions, when little charge is being drawn from V_{DD}, the loop filter stops activation of the power FETs which are the largest source of loss. Only one out of N cycles generates an "on" pulse, where N can be as high as 100 cycles.

The entire front-end is digital, which includes all the circuits starting from V_{DD} up to the generation of f_{ERR}. When $-3 \leq f_{ERR} \leq 0$, these are the only circuits actively operating and dissipating power. By taking their variable delay over voltage into account during the design of the loop, they can all be powered from V_{DD}, instead of V_{BAT}. Thus, the power of these circuits, which are continuously running, scales with the current V_{DD} operating point, so that at low voltage, their power dissipation becomes insignificant.

To improve efficiency while the buck converter is actively operating, the power FETs are comprised of multiple parallel FETs. Then, the actual FET size is dynamically varied to minimize loss over the range of operating conditions [7.3].

DC-DC Voltage Conversion

The combination of these techniques provides an efficiency of 80-95% while the processor is actively operating over the range of voltage and current load, and has negligible power loss while the processor is idling. Chapter 8 describes the energy efficiency of a prototype implementation in further detail.

7.5.5 New Performance Metrics

In addition to the supply ripple and conversion efficiency performance metrics of a standard voltage regulator, the DVS converter introduces two new performance metrics: transition time and transition energy. For a large voltage change ($V_{DD1} \rightarrow V_{DD2}$), the transition time is:

$$t_{TRAN} \approx \frac{2 \cdot C_{DD}}{I_{MAX}} \cdot |V_{DD2} - V_{DD1}| \qquad \text{(EQ 7.53)}$$

where I_{MAX} is the maximum output current of the converter, and the factor of 2 exists because the current pulses are triangular. In practice, t_{TRAN} will be slightly longer for a low-to-high voltage transition because the actual current charging C_{DD} is $I_{MAX} - I_{DD}(V_{DD})$. The energy consumed during this transition is:

$$E_{TRAN} = (1 - \eta) \cdot C_{DD} \cdot |V_{DD1}^2 - V_{DD2}^2| \qquad \text{(EQ 7.54)}$$

Since both transition time and transition energy are proportional to C_{DD}, minimizing C_{DD} yields a faster and more energy-efficient voltage converter.

To gauge how the transition energy impacts the overall system energy consumption, it is more intuitive to compare the power dissipation which factors in the frequency of voltage transitions and level of processor performance. Given a frequency, f_{VDD}, at which the system makes voltage transitions, the transition power dissipation is:

$$P_{TRAN} = E_{TRAN} \cdot f_{VDD} = (1 - \eta) \cdot C_{DD} \cdot |V_{DD1}^2 - V_{DD2}^2| \cdot f_{VDD} \qquad \text{(EQ 7.55)}$$

Figure 7.23 demonstrates how transition time (t_{TRAN}) and transition power dissipation (P_{TRAN}) vary with C_{DD} for the maximum 1.2-3.8V voltage transition of the prototype system, which has $I_{MAX} = 1A$, $\eta = 90\%$.

Dynamic Voltage Conversion

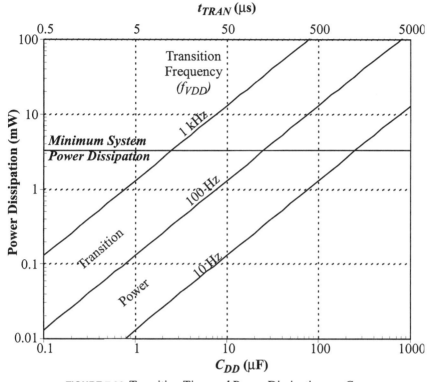

FIGURE 7.23 Transition Time and Power Dissipation vs. C_{DD}.

P_{TRAN} is shown for three different values of f_{VDD}. Also plotted is the minimum prototype system power dissipation not including P_{TRAN}, and sets the threshold below which P_{TRAN} should remain so that it does not dominate the total system power dissipation. A typical C_{DD} value for low-voltage/low-power voltage regulators is 100μF. This gives a t_{TRAN} in excess of 500μs which precludes any real-time control or fast interrupt response time, and only allows very coarse speed control. For this value of C_{DD}, an f_{VDD} on the order of a context switch (30-100Hz) will cause the transition power to dominate the system power (55-80% of the total power).

Thus, existing voltage regulators make very poor voltage converters due to their large C_{DD}, which needs to be reduced by at least 10x. Using the converter loop, combined with the hybrid PWM/PFM algorithm, allowed a dynamic voltage regulator to be designed which maintains good conversion efficiency at much lower values of C_{DD}.

7.5.6 Limits to Reducing C_{DD}

Decreasing C_{DD} reduces transition time, and by doing so increases the speed at which the voltage changes, dV_{DD}/dt. CMOS circuits can operate with a varying V_{DD}, but only up to a point, which is process dependent.

Decreasing C_{DD} also increases supply ripple, which in turn increases processor energy consumption as shown in Figure 7.24.

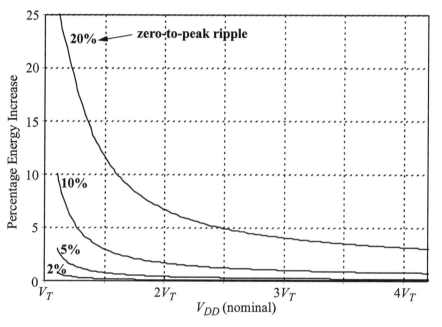

FIGURE 7.24 Energy Loss Due to Voltage Supply Ripple.

The increase is moderate at high V_{DD}, but begins to increase as V_{DD} approaches V_T because the negative ripple slows down the processor so much that most of the computation is performed during the positive ripple, which decreases energy efficiency. For values of supply ripple above 10%, the processor can still operate properly, but the increased energy consumption of the processor outweighs the decreased transition energy consumption, degrading overall system energy-efficiency.

Dynamic Voltage Conversion

Loop stability is another limitation on reducing capacitance. As described in Section 7.5.2, the dominant pole in the system is inversely proportional to C_{DD}. As C_{DD} is reduced the pole frequency increases. As the pole approaches the sampling frequency, interaction with higher-order poles will eventually make the system unstable.

The third limitation is that low-voltage conversion efficiency scales down with C_{DD}. Since the DVS processor will ideally be operating most of the time at low voltage, it is important to maintain reasonable low-voltage conversion efficiency.

Increasing the converter sampling frequency will reduce the supply ripple and increase the pole frequency due to the sample delay. Thus, these two limits are not fixed, but can be varied. However, increasing the sampling frequency has two negative side-effects. First, low-load converter efficiency will decrease because the converter loop will need to be activated more frequently to maintain the same voltage. Second, the f_{CLK} quantization error will increase. These side-effects may be mitigated with a variable sampling frequency that adapts to the system power requirements (e.g. V_{DD} and I_{DD}).

The maximum dV_{DD}/dt at which the circuits will still operate properly is a hard constraint because system failure can be induced, but occurs for a much smaller C_{DD} than the supply ripple and stability constraints. Low-voltage conversion efficiency is a soft-constraint, but cannot be improved by adjusting the converter sampling frequency.

7.5.7 Optimizing C_{DD} in the Prototype System

For the prototype system, a value of 5µF was chosen for C_{DD}. The limiting factor for not reducing it further was the low-voltage conversion efficiency. Table 7.1 lists the key converter performance parameters for both the typical value of C_{DD} (100µF) and the optimized value (5µF). The top four parameters were optimized given the three bottom hard constraints. The optimized value maintains the constraints placed on supply ripple, the dominate pole frequency, and dV_{DD}/dt, while minimizing E_{TRAN} and t_{TRAN} and maintaining good high voltage (3.3V) and low-voltage (1.2V) conversion efficiency. There is still plenty of margin for the hard constraints, which would allow for an even smaller C_{DD} if the converter loop could be redesigned to compensate for

DC-DC Voltage Conversion

the reduction in low-voltage conversion efficiency, and continue to maintain a reasonable value (> 80%).

TABLE 7.1 Converter Performance Parameters

Parameter	Constraint	$C_{DD} = 100\mu F$	$C_{DD} = 5\mu F$
E_{TRAN}	minimize	130 µJ/transition	6.5 µJ/transition
t_{TRAN}	minimize	~520 µs	~26 µs
η (3.3V)	maximize	> 95%	92%
η (1.2V)	maximize	> 95%	84%
ripple	< 10%	< 1%	2%
dom. pole	< 100 kHz	400 Hz	7 kHz
dV_{DD}/dt	< 5 V/µs	0.01 V/µs	0.2 V/µs

References

[7.1] J. Kassakian, M. Schlecht, and G. Verghese, *Principles of Power Electronics*, Addison-Wesley, 1991.

[7.2] A. Stratakos, S. Sanders, and R.W. Brodersen, "A Low-voltage CMOS DC-DC Converter for Portable Battery-operated Systems", *Proceedings of the Twenty-Fifth IEEE Power Electronics Specialist Conference*, June 1994, pp. 619-626.

[7.3] A. Stratakos, *High-Efficiency, Low-Voltage DC-DC Conversion for Portable Applications*, Ph.D. Thesis, University of California, Berkeley, 1998.

[7.4] D. Aebischer, et. al., "A 2.1-MHz Crystal Oscillator Time Base with a Current Consumption under 500nA", *IEEE Journal of Solid State Circuits*, Vol. 32, No. 7, Jul. 1997, pp. 999-1005.

[7.5] J. Montanaro, et. al., "A 160-MHz 32-b 0.5-W CMOS RISC Microprocessor", *IEEE Journal of Solid State Circuits*, Vol. 31, No. 11, Nov. 1996, pp. 1703-14.

CHAPTER 8 DC-DC Converter IC for DVS

In a DVS microprocessor subsystem, the processor core and surrounding peripherals are run from a dynamically scaled voltage supply, enabling up to a 10x improvement in average energy per operation. This section describes the implementation of a prototype dynamic DC-DC converter for application in a DVS system along with measured results.

8.1 System and Algorithm Description

Figure 8.1 shows a block diagram of the dynamic DC-DC converter prototype IC in its DVS application. The desired frequency, f_{DES}, is determined by the process scheduler and then is provided to the hardware through the 7-bit digital word, M, given in Equation 7.49. The DVS loop then sets the processor clock (a buffered version of f_{VCO}) to f_{DES}, by adjusting the supply voltage, V_{DD}, as rapidly as possible.

To increase efficiency the dynamic DC-DC converter is designed to operate in a discontinuous mode using a synchronous PWM-PFM control scheme. By exploiting the 4 MHz DVS system clock and using low voltage digital control bootstrapped from the converter output, the controller achieves low static power dissipation which scales together with the load. Pulse-width modulation commands the quantity of charge

DC-DC Converter IC for DVS

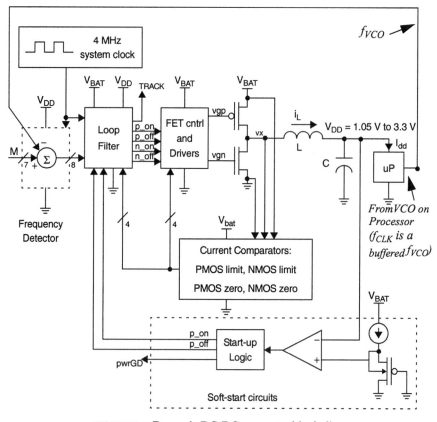

FIGURE 8.1 Dynamic DC-DC converter block diagram.

delivered during each PFM pulse through the controlled power FET conduction interval. A pulse skipping algorithm modulates the pulse frequency, maintaining acceptable conversion efficiency over the dynamic range of the converter.

A system timing diagram is shown in Figure 8.2. The frequency detector generates an 8-bit digital representation of the frequency error, f_{ERR}, every 1 μs. The loop filter samples f_{ERR} on the following falling edge of clk4. In the first cycle of Figure 8.2, $f_{ERR}= -1$, and the converter idles until the next sampling instant. During this interval, the processor discharges V_{dd}, causing a corresponding decrease in f_{VCO}. When the sampled $f_{ERR}> 0$, the loop filter translates f_{ERR} into an update command for the DC-DC converter. A PFM pulse is initiated by the PMOS power FET, and the power

NMOS functions as a synchronous rectifier, turned off by the NMOS zero current comparator when i_{dsN} decays to zero. The cycle then repeats.

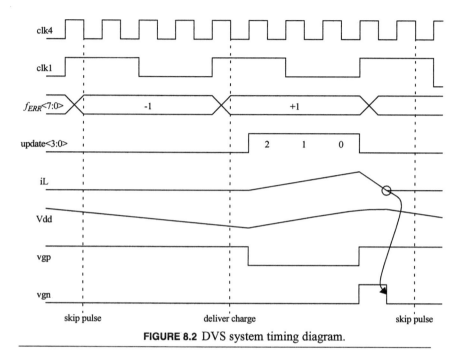

FIGURE 8.2 DVS system timing diagram.

8.1.1 PWM Control

The pulse-width modulation algorithm contains proportional and feedforward terms (Figure 8.3). A power FET conduction interval, T_{on}, is the controlled variable. For a quantized frequency error:

$$f_{ERR} = \text{floor}\left(\frac{f_{DES} - f_{VCO}}{1 \text{ MHz}}\right) \quad \text{(EQ 8.1)}$$

the controlled conduction interval is:

$$T_{on} = (250 \text{ ns}) \cdot (\text{feedforward} + \text{gain} \cdot f_{err}) \quad \text{(EQ 8.2)}$$

In Equation 8.2, the T_{on} LSB is 250 ns, equal to one cycle of the 4 MHz DVS system clock. The feedforward term is chosen as a function of M to sustain full load current or to consume a 2% peak-to-peak output voltage ripple budget.

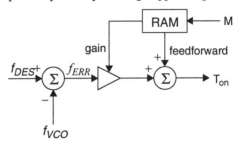

FIGURE 8.3 PWM block diagram.

The transfer function is two-sided (Figure 8.4). For $f_{ERR} < 0$, $T_{on} < 0$ and the converter removes excess charge from its output capacitor. The PFM pulse is initiated by the NMOS power FET, $T_{nmos} = |T_{on}|$, and the power PMOS is operated as a synchronous rectifier. For $f_{ERR} > 0$, the converter delivers charge to the output via a PFM pulse initiated by the PMOS power FET.

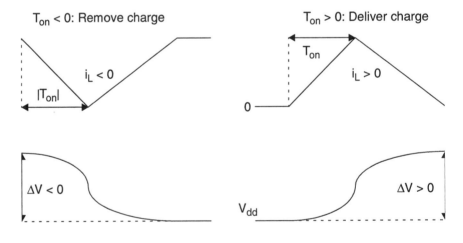

FIGURE 8.4 Charge removal and delivery.

System and Algorithm Description

Current limiting is included to protect the power FETs and external filter elements during large signal tracking transitions. The magnitude of peak positive and negative inductor currents are limited to 1 A.

8.1.2 PFM Control

Pulse frequency modulation ensures that the converter switches only when necessary, conserving power at low output voltage and light load. The pulse-skipping algorithm is simple: For $-3 \leq f_{err} < 0$, the converter idles, allowing the processor to discharge V_{dd}, decreasing f_{VCO}. For $f_{err} \geq 0$ or $f_{err} < -3$, charge is delivered to or removed from the output according to the PWM algorithm of Equation 8.2.

Figure 8.5 summarizes the transfer function of the hybrid PWM-PFM controller.

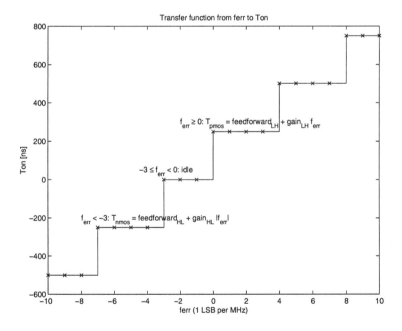

FIGURE 8.5 PWM-PFM transfer function from f_{ERR} to T_{on}.

PWM parameters $gain_{LH}$, $gain_{HL}$, feedforward$_{LH}$ and feedforward$_{HL}$ are chosen as a function of the desired frequency, M (Equation 7.49). In Figure 8.5, f_{DES} = 24 MHz

DC-DC Converter IC for DVS

(M = 24), gain_{LH} = gain_{HL} = 1/4 LSB per MHz, feedforward_{LH} = 1 LSB, and feedforward_{HL} = 0.

8.1.3 Start-Up

A reliable start-up mechanism is required to enable bootstrapped operation of the digital controller. Figure 8.6 shows a block diagram of the approach. At power-on, V_{dd} = 0, and the soft-start controller commands the DC-DC converter. A simple synchronous PFM scheme, with a constant 500 ns on-time, is used to ramp the output voltage. Once the output voltage exceeds a weak PMOS $V_{GS} \approx 1.2$ V , the pwrGD flag is raised, and the DVS controller assumes command of the converter, initialized with M = 24. When 21 MHz < f_{VCO} < 27 MHz, the TRACK signal falls, indicating successful frequency regulation.

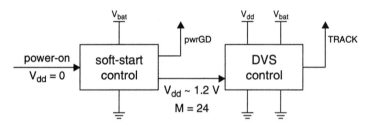

FIGURE 8.6 Start-up algorithm.

8.1.4 System Simulation Results

The processor is expected to achieve an energy per operation of 4.5 nJ at the 3.3 V, 100 MIPS operating point, yielding an average full-load current of 135 mA. The energy per operation scales with voltage as described in Equation 2.7. This data, together with the modeled ring oscillator performance of Figure 8.7, is used to generate a curve of nominally expected processor full-load current versus throughput (Figure 8.8)

Using this processor model the control system has been verified using Matlab simulation. Figure 8.9 shows the simulated tracking performance with V_{bat} = 3.6 V, L = 3.5 µH, and C = 4.7 µF. The large-signal 12 MHz to 90 MHz tracking transition settles within 20 µs.

External Component Selection

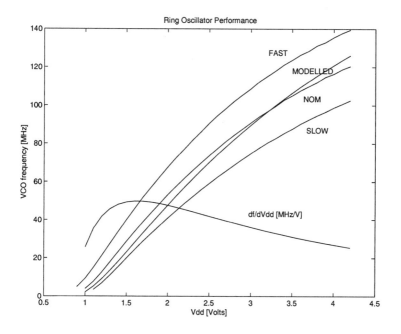

FIGURE 8.7 Simulated and modeled ring oscillator performance.

Figure 8.10 shows regulation at the desired clock rates of 26 MHz and 95 MHz. The DC-DC converter pulse width and pulse frequency are reduced at the lower output frequency. Output voltage ripple is kept below 2% at 26 MHz.

8.2 External Component Selection

Tracking and regulation metric trade-offs through filter element sizing have been examined in Section 7.5. Here, minimization of output capacitance for superior tracking metrics, with acceptable output voltage ripple and low-voltage efficiency, is the primary design objective. Q_L, L, and C are chosen according to Equation 7.23, Equation 7.25, and Equation 7.26 to sustain full load current in a 4 μs minimum repetition period with acceptable output voltage ripple.

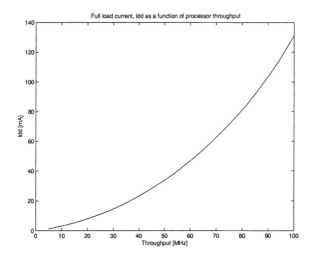

FIGURE 8.8 Expected processor full-load current.

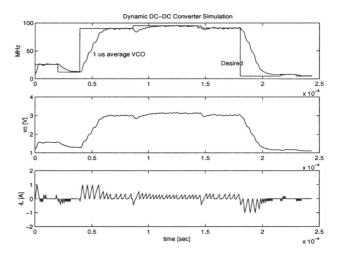

FIGURE 8.9 Simulated tracking performance.

External Component Selection

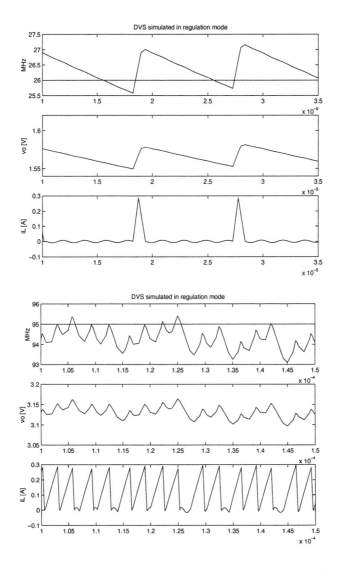

FIGURE 8.10 Simulated regulation waveforms at 26 MHz (top) and 95 MHz (bottom).

DC-DC Converter IC for DVS

L = 3.5 µH and C = 4.7 µF are selected as a reasonable compromise between tracking and regulation metrics. A fourfold improvement in tracking time and a sixfold improvement in tracking energy were found over previous dynamic DC-DC converters *[8.1], [8.2], [8.3]*. Power train and output voltage ripple losses are kept below 4% at the low throughput corner. Figure 8.11 shows the charge delivered per PFM pulse, the PMOS and NMOS conduction intervals, the output voltage ripple, and the normalized regulation energy dissipation as a function of processor throughput for L = 3.5 µH, C = 4.7 µF, and V_{bat} = 3.6 V.

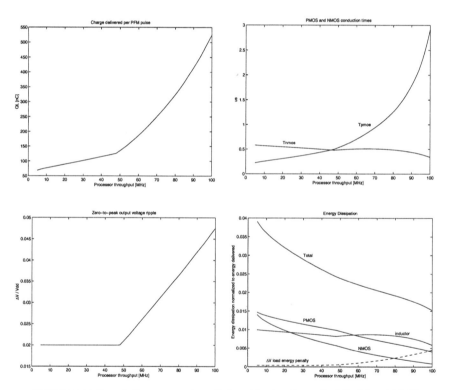

FIGURE 8.11 Regulation parameters. Pulse-skipping is applied for M < 48 MHz.

8.3 Frequency Detector

Figure 8.12 shows the frequency detector, which generates a digital representation of the VCO frequency error averaged over a 1 μs period. The operating system's process scheduler determines the desired processor throughput, requesting an integer multiple, M, of 1 MHz.

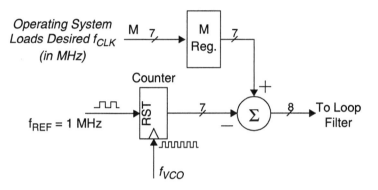

FIGURE 8.12 Digital frequency detector.

A seven-bit counter clocks rising edges from the VCO output frequency, f_{VCO}. The reference frequency, $f_{REF} = 1$ MHz, which is derived from the 4 MHz DVS system clock, asserts the asynchronous reset of the counter, resetting its output to zero every 1 μs. Just prior to the asynchronous reset, the output of the counter is given by:

$$\text{count}(k) = \text{floor}\left(\frac{f_{VCO}}{1 \text{ MHz}}\right) + \text{remainder}(k-1) \quad \text{(EQ 8.3)}$$

where remainder(k-1) is the remainder of the truncation of Equation 8.3 performed in cycle (k-1).

This output is latched and subtracted from the 7-bit digital representation of the desired frequency, M, yielding an 8-bit two's complement digital error signal:

$$f_{err}(k) = M - \text{count}(k) \quad \text{(EQ 8.4)}$$

which is proportional to the frequency error, averaged over cycle k, with an LSB of 1 MHz.

The frequency detector introduces a cycle-by-cycle quantization error which becomes increasingly significant at lower processor throughputs. At the minimum throughput of 5 MHz, cycle-by-cycle quantization error can be as high as 20%. However, as illustrated by Figure 8.13, while the error is truncated every 1 µs, the remainder of the error accumulates in the frequency detector, forcing the average quantization error to zero. Thus, quantization contributes no DC offset to V_{dd} and f_{VCO}, but does introduce additional AC ripple.

The frequency detector continuously evaluates, regardless of the converter's loading conditions, and therefore, consumes static power. So that its energy consumption scales at lower output voltages, it is operated from the voltage scaled supply, V_{dd}. The effective capacitance includes a 7-bit counter switching at the VCO output frequency, a 2-bit clock divider switching at 4 MHz, and a 7-bit register and 8-bit adder switching at 1 MHz. The average power dissipation is given by:

$$P_{\text{FreqDetect}} = (1.1 \text{ pF}) \cdot f_{VCO} \cdot V_{dd}^2 + (3.6 \text{ pF}) \cdot (1 \text{ MHz}) \cdot V_{dd}^2 \quad \text{(EQ 8.5)}$$

contributing 10 µW at the 5 MHz, 1.05 V operating point, and 1.2 mW at the 100 MHz, 3.3 V operating point.

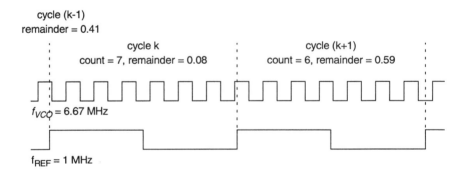

FIGURE 8.13 Quantization error in the frequency detector.

The VCO output is driven from the processor to the dynamic DC-DC converter IC. If swung rail-to-rail, its power consumption might prove to be the dominant contributor to overall dissipation in the DVS loop. At the low throughput corner:

$$P_{VCO} = (20\text{ pF}) \cdot (1.05\text{ V})^2 \cdot (5\text{ MHz}) = 110\text{ }\mu W \quad \text{(EQ 8.6)}$$

At the high-throughput corner:

$$P_{VCO} = (20\text{ pF}) \cdot (3.3\text{ V})^2 \cdot (100\text{ MHz}) = 22\text{ mW} \quad \text{(EQ 8.7)}$$

If, instead, the 20 pF of parasitic capacitance is driven by the low-swing I/O transmitter of Section 4.3 powered by the 200 mV output of a DC-DC converter, the total power dissipated in driving the inter-chip capacitance is significantly reduced.

$$P_{VCO} = (20\text{ pF}) \cdot (0.2\text{ V})^2 \cdot (5\text{ MHz}) = 4\text{ }\mu W \quad \text{(EQ 8.8)}$$

at the low-throughput corner, and:

$$P_{VCO} = (20\text{ pF}) \cdot (0.2\text{ V})^2 \cdot (100\text{ MHz}) = 80\text{ }\mu W \quad \text{(EQ 8.9)}$$

at the high-throughput corner.

The dynamic DC-DC converter includes a receiving pad to decode the incoming 200 mV signal. A description of the receiver can be found in Section 4.3. Its power consumption is given by:

$$P_{receiver} = (15\text{ }\mu A) \cdot V_{bat} + (0.9\text{ pF}) \cdot V_{dd}^2 \cdot f_{VCO} \quad \text{(EQ 8.10)}$$

yielding 45 µW at 5 MHz, 1.05 V, and 1.0 mW at 100 MHz, 3.3 V. The total power savings effected by the low-swing VCO transceiver is 1.8x at the low throughput corner, and 20x at the high throughput corner.

DC-DC Converter IC for DVS

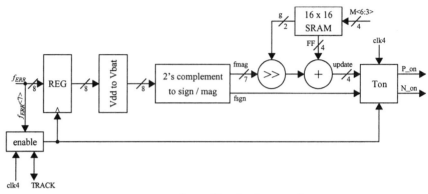

FIGURE 8.14 Loop filter implementation.

8.3.1 Loop Filter

The loop filter translates f_{ERR} into an update command for the DC-DC converter. It implements the pulse-width modulation and pulse-skipping algorithms. It is responsible for hand-off between regulation and tracking modes.

Tracking mode is initiated by a new frequency request from the process scheduler. In tracking mode, the converter is capable of slewing its output up and down. When $f_{err} > 0$, the VCO frequency is too low, and the converter is commanded to deliver charge to the output capacitor. The PMOS device initiates the PFM pulse, T_{pmos} is the controlled variable, and the NMOS power FET acts as a synchronous rectifier. When $f_{err} < 0$, the VCO frequency is too high, and the converter is commanded to remove charge from the output capacitor. The NMOS device initiates the PFM pulse, T_{nmos} is the controlled variable, and the PMOS power FET acts as a synchronous rectifier. When $-4 < f_{err} < 4$, control is handed to regulation mode.

In regulation mode, the converter can only deliver charge to the output capacitor, it cannot remove it. When $f_{err} \geq 0$, a PFM pulse is initiated by the power PMOS device. When $f_{err} < 0$, the converter idles and the loop filter continues to monitor the frequency error until $f_{err} \geq 0$.

Figure 8.14 shows a block diagram of the loop filter implementation. f_{ERR} swings at V_{dd}; all other signals are driven at V_{bat}. The "enable" block implements the pulse-

Frequency Detector

skipping function, clocking f_{ERR} on the falling edge of clk4 under the following set of conditions:

1) Neither power FET is conducting, and
2) TRACK is high, or
3) TRACK is low and f_{ERR}<7> is high

The 8-bit two's complement f_{ERR} is level-shifted to V_{bat} and converted to an 8-bit sign / magnitude representation. In tracking mode, f_{sgn} determines which power FET is controlled.

The PWM algorithm is given in Equation 8.2. An intermediate variable, update, is a 4-bit unsigned word:

$$\text{update} = \text{FF} + 2^{-g} \cdot f_{mag} \qquad \text{(EQ 8.11)}$$

which stores T_{on} in LSB. The loop filter saturates at update = 15, constraining the maximum on-time to 3.75 μs. Feedforward and gain terms are set as a function of the four MSB's of the desired frequency, M. Unique values of FF and g are chosen for low-to-high and high-to-low tracking transitions.

The "Ton" block negotiates power FET sequencing and converts update into a controlled conduction interval:

$$T_{on} = \text{update} \cdot 250 \text{ ns} \qquad \text{(EQ 8.12)}$$

The loop filter consumes no static power: It switches only during active PFM pulses. The energy dissipated per DC-DC converter switching event is data dependent, but for high-level energy budgeting, it is approximated by:

$$E_{filter} = (1.7 \text{ pF}) \cdot V_{dd}^2 + (9.2 \text{ pF}) \cdot V_{bat}^2 \qquad \text{(EQ 8.13)}$$

which equals 120 pJ (0.2%) at the low throughput corner and 138 pJ (negligible) at the high throughput corner.

8.4 Current Comparators

The prototype converter uses four sets of offset-cancelled comparators for zero-current detection and current limiting in the power transistors. The comparator topology is shown in Figure 8.15 [8.4]. Two input-offset cancelled differential amplifier stages

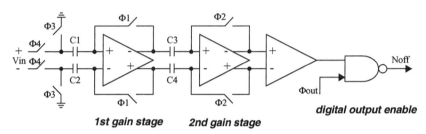

FIGURE 8.15 i_{NMOS} comparator topology.

form the main preamplifier. A high gain differential to single-ended amplifier and a nand gate convert the output to full-swing digital levels. In reset mode, the bias to the amplifiers is disabled, switches phased $\Phi1$, $\Phi2$, and $\Phi3$ are closed, and switches phased $\Phi4$ are open. To conserve static power, the master control does not enable the comparator bias until exactly 1 ms after the power NMOS device is gated. In the succeeding 250 ns, the preamplifier offset is stored on the interstage coupling capacitors. The input capacitors also serve to level-shift the inputs, extending the input common-mode range below ground.).

8.4.1 PMOS current limit

The PMOS current limit protects the power FETs and external filter elements during large signal tracking transitions. The peak conducted PMOS current is limited to 0.5 A or 1.0 A in tracking mode.

The circuit implementation is shown in Figure 8.16. It consists of one offset-cancelled comparator, a x1 reference FET, identically matched to the xN power FET, and a known current i_{REF}. The comparator begins to switch when inductor current, i_L, conducted through the PMOS power FET induces a source-to-drain voltage drop greater than that induced by i_{REF} flowing through the reference FET. The accuracy of the comparator trip point:

Current Comparators

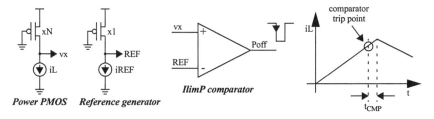

FIGURE 8.16 PMOS current limit implementation.

$$i_L = i_{REF} \cdot N \quad \text{(EQ 8.14)}$$

is determined primarily by the control on the absolute value of i_{REF}, and the matching of the x1 reference FET to the xN PMOS power FET.

This circuit is activated only during tracking PFM pulses which are initiated by the PMOS device. It includes a strobed bias network and gated clocks for low-power. It dissipates no static power during regulation mode.

8.4.2 NMOS current limit

The NMOS current limit is nearly identical to the PMOS current limit of Figure 8.16. It is activated only during tracking PFM pulses which are initiated by the NMOS device, and dissipates no static power during regulation mode. Peak negative NMOS current is limited to -0.5 A or -1.0 A.

8.4.3 NMOS zero-current detection

The i_{NMOS} comparator implementation commands the turn-off transition of the NMOS synchronous rectifier when i_{dsN} crosses zero from above.

The equivalent input-referred offset voltage were $V_{os} = 0.5$ mV and delay $t_{cmp} \sim 50$ ns of the comparator resulted in a worst-case NMOS turn-off current error of:

$$I_\varepsilon = \frac{0.5 \text{ mV}}{160 \text{ m}\Omega} + (50 \text{ ns}) \cdot \frac{1.05 \text{ V}}{3.5 \text{ }\mu H} = 3.1 \text{ mA} + 15.0 \text{ mA} \quad \text{(EQ 8.15)}$$

and

$$I_\varepsilon = \frac{0.5 \text{ mV}}{40 \text{ m}\Omega} + (50 \text{ ns}) \cdot \frac{3.3 \text{ V}}{3.5 \text{ }\mu H} = 12.5 \text{ mA} + 47.1 \text{ mA} \quad \text{(EQ 8.16)}$$

for the low and high throughput operating points. This translates to worst-case energy dissipation penalties of 0.57 nJ (0.8%) and 6.2 nJ (0.4%), respectively. In an effort to reduce these dissipation penalties, an integral feedback loop is used to null the comparator, logic, and power FET gate-drive delays. Figure 8.17 describes the approach.

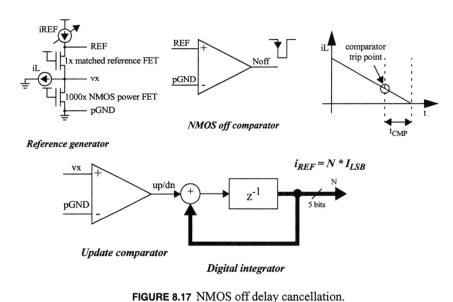

FIGURE 8.17 NMOS off delay cancellation.

The circuit implementation includes two identical offset-cancelled comparators. The NMOS off comparator commands the power NMOS turn-off transition. The update comparator monitors the results and adapts the NMOS off trip point to null its delay.

The NMOS off comparator begins to switch when

$$v_{REF} = v_{pGND} \quad \text{(EQ 8.17)}$$

where pGND is a Kelvin connection to the power NMOS source terminal. The reference generator includes a matched reference FET and a digitally-programmable current source, i_{REF}, so that the trip point of the comparator is given by:

$$i_L = i_{REF} \cdot \frac{W_{NMOS}}{W_{REF}} = 1000 \cdot i_{REF} \qquad \text{(EQ 8.18)}$$

Proper adjustment of i_{REF} is ensured by the integral feedback loop. A digital integration scheme is selected to allow maintenance of state without static power dissipation. The effective LSB is $i_{NMOS} \sim 2$ mA.

Gated clocks and strobed biasing are used to eliminate static power. The comparators are enabled by the power PMOS turn-on – during positive PFM pulses only – and are disabled 125 ns after NMOS turn-off. The reference generator, with 0 to 62 µA of static current, is enabled 125 ns after NMOS turn-on, and is disabled at NMOS turn-off. The overall energy dissipated per NMOS off event is given by:

$$E = \frac{1}{2}LI_\varepsilon^2 + (8.0 \text{ pF})V_{bat}^2 + V_{bat} \cdot ((310 \text{ μA})(T_p + T_n + 125 \text{ ns}) + (30 \text{ μA})T_n) \qquad \text{(EQ 8.19)}$$

Equation 8.19 includes the energy dissipation penalty associated with early or late NMOS turn-off, and assumes $i_{REF} = 30$ µA. For $V_{bat} = 3.6$ V, E = 1.2 nJ (1.7%) at the low throughput corner. Here, it is interesting to note that the adaptive timing control actually costs 60 pJ of additional dissipation. At the high throughput corner, E = 4.1 nJ (0.2%), and the adaptive timing control conserves 4.0 nJ.

8.4.4 PMOS zero-current detection

The PMOS off comparator is nearly identical to the NMOS off comparator. It commands the turn-off transition of the PMOS synchronous rectifier when i_{dsP} crosses zero from below. It includes an adaptive timing control loop to null comparator, logic, and power FET gate-drive delays.

The comparators are enabled by the power NMOS turn-on – during negative PFM pulses only – and are disabled 125 ns after PMOS turn-off. The bias is never enabled during regulation mode. Strobed biasing and gated clocks assure that it dissipates no static power.

DC-DC Converter IC for DVS

8.5 Power FETs

The integrated power FETs are binary weighted, with two control bits each for independent dynamic NMOS and PMOS sizing. The NMOS and PMOS gate-width LSB's are 10 mm and 20 mm, respectively. The minimum drawn channel length of 0.6 μm is used.

Figure 8.18 shows the power FETs, gate-drive, and dynamic transistor sizing modules. The FETs are dynamically sized versus requested throughput, M, a-priori, with appropriate control bits Wp0, Wp1, Wn0, Wn1 stored in RAM. Switching and gate-drive loss are traded with conduction loss at each operating point with the total FET energy dissipation given by:

$$E_{diss} = \frac{1}{3} \cdot I_p^2 \cdot \left(T_p \cdot \frac{R_p}{W_p} + T_n \cdot \frac{R_n}{W_n} \right) + V_{bat}^2 \cdot (C_{overhead} + W_p \cdot C_p + W_n \cdot C_n) \quad \text{(EQ 8.20)}$$

where subscripts p and n indicate contributions due to PMOS and NMOS power transistors; I_{peak} is the peak PFM pulse current, found from Equation 7.20; W is the gate-width in LSB; T is the conduction time interval, found from Equation 7.21; R is the effective channel resistance of an LSB, listed in Table 8.1 ; and C is the effective switched capacitance of an LSB:

$$C = C_{gd} + C_{gs} \quad \text{(EQ 8.21)}$$

C_p = 49 pF and C_n = 32 pF also accounts for dissipation in the gate drive.

$C_{overhead}$ is the overhead capacitance, equal to:

$$C_{overhead} = 3C_{gdp} + 3C_{gdn} + 3C_{dbp} + 3C_{dbn} + C_x = 120 \text{ pF} \quad \text{(EQ 8.22)}$$

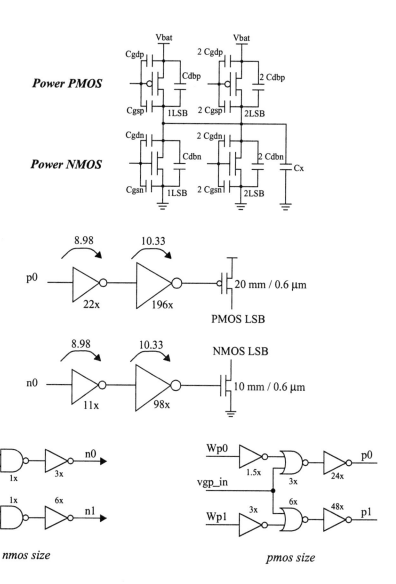

FIGURE 8.18 Power FETs, gate-drive, and dynamic sizing module.

DC-DC Converter IC for DVS

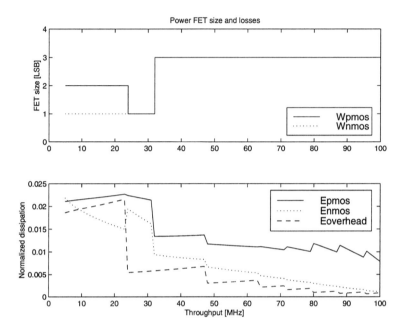

FIGURE 8.19 Prototype power FET size and losses.

Figure 8.19 shows the gate-widths and expected energy dissipation for the prototype IC implementation.

TABLE 8.1 Simulated power FET LSB channel resistance.

	Rp	Rn
slow, 3.0 V	440 mΩ	224 mΩ
nom, 3.6 V	343 mΩ	189 mΩ
fast, 4.2 V	289 mΩ	168 mΩ

8.6 Efficiency Simulations

Figure 8.20 plots the expected converter efficiency versus throughput at full-load and at one-quarter-load. .

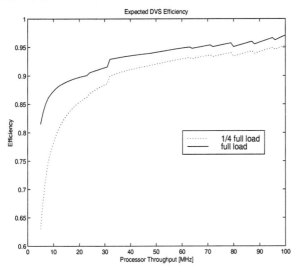

FIGURE 8.20 Efficiency vs. processor throughput at heavy and medium loads.

The mechanisms of loss in the DVS system are summarized in Table 8.2 .

TABLE 8.2 Mechanisms of loss in the DVS system.

Mechanism of Loss	Equation or Source	5 MHz 1.05 V 0 mA	5 MHz 1.05 V 1.2 mA	100 MHz 3.3 V 0 mA	100 MHz 3.3 V 135 mA
PROCESSOR					
Processor	Figure 8.8	0	1.26 mW	0	446.0 mW
VCO	Simulated result	6.3 µW	6.3 µW	0.98 mW	0.98 mW
Low-swing VCO interconnect	Equation 8.8 and Equation 8.9	4.0 µW	4.0 µW	80.0 µW	80.0 µW

TABLE 8.2 Mechanisms of loss in the DVS system.

Mechanism of Loss	Equation or Source	5 MHz 1.05 V 0 mA	5 MHz 1.05 V 1.2 mA	100 MHz 3.3 V 0 mA	100 MHz 3.3 V 135 mA
TOTAL LOAD	uP + VCO + transmitter	10.3 µW	1.27 mW	1.06 mW	447.0 mW
DC-DC Conv					
Master Bias	20 µA static current from V_{bat}	72 µW	72 µW	72 µW	72 µW
VCO receiver	Equation 8.10	45 µW	45 µW	1.0 mW	1.0 mW
Freq Detect	Equation 8.5	10 µW	10 µW	1.2 mW	1.2 mW
Loop Filter	Equation 8.13	0	1.9 µW	0	35.1 µW
NMOS off	Equation 8.19	0.2 µW	18.6 µW	2.7 µW	1.1 mW
FET control	C_{eff} = 1.6 pF at V_{bat} per PFM pulse	0	0.3 µW	0	5.7 µW
Power FETs	Equation 8.20	0.6 µW	74.7 µW	9.6 µW	4.1 mW
L	Equation 7.29; $R_{L(dc)}$ = 0.09 Ω; $R_{L(ac)}$ = 0.3 Ω	0.1 µW	14.7 µW	6.0 µW	2.5 mW
C	Equation 7.29 R_{esr} = 0.08 Ω	0.1 µW	11.7 µW	4.7 µW	0.6 mW
Stray inductance	Equation 7.45 L_s = 9.0 nH	0	2.3 µW	0.2 µW	96.0 µW
Series resistance	Equation 7.41; $R_{s(pmos)}$ = 18 mΩ; $R_{s(nmos)}$ = 18 mΩ	0	5.2 µW	2.1 µW	0.9 mW
TOTAL LOSS	Σ (All converter losses)	128.0 µW	256.4 µW	2.30 mW	11.6 mW
SYSTEM DISSIPATION		138.3 µW	1.53 mW	3.36 mW	458.6 mW
EFFICIENCY		-	83.8%	-	97.5%

All losses in the power train, controller, and processor load are considered. The DVS system is expected to dissipate 138 µW and 3.4 mW of static power at the low throughput and high throughput corners, with the converter consuming the majority of the power. Here, the primary mechanisms of dissipation include the processor VCO, and the VCO receiver, frequency detector, and master bias of the DC-DC converter. Considering all losses in the processor and converter at full-load, the system energy per operation is expected to be 0.3 nJ/instruction at 5 MIPS and 1.05 V, and 4.6 nJ/instruction at 100 MIPS and 3.3 V.

8.7 Measured Results

The prototype converter was fabricated in a 0.6 micron, single poly, triple metal CMOS process. Figure 8.21 shows the IC layout, with die dimensions of 1.68 mm x 3.41 mm. The power section includes 1.6 nF of integrated bypass capacitance tuned to $\tau_{RC} = 2.6$ ns. Considerable die area is devoted to the six offset-cancelled comparators, whose offset storage capacitors are implemented as metal1-metal2-metal3 sandwiches. Separate power FET, high-voltage digital, low-voltage digital and analog supplies are maintained for isolation and power characterization. The IC is assembled in a 68 J-lead ceramic chip carrier, and mounted to the printed circuit board in a through-hole socket.

The IC tracks frequency requests in the μs to tens of μs time scale, with 80% to 90% full-load efficiencies over the full 5 MHz to 100 MHz dynamic range. The following subsections detail the results

8.7.1 Start-UP, Tracking Performance and Current Limiting

In Figure 8.22, is shown the soft-start transient from $V_{dd} = 0$ to $V_{dd} = 1.2$ V is captured . with Figure 8.23 showing a full-scale 5 MHz to 100 MHz tracking transition with $V_{bat} = 6.0$ V, at medium load, and with a 1.0 A current limit. The low-to-high tracking time of 23.5 μs is slew limited by the forward PMOS current limit. The high-to-low tracking transition is slower by design and measured to be 44.0 μs. The -1.1 A reverse NMOS current limit slew limits the early portion of the output voltage excur-

DC-DC Converter IC for DVS

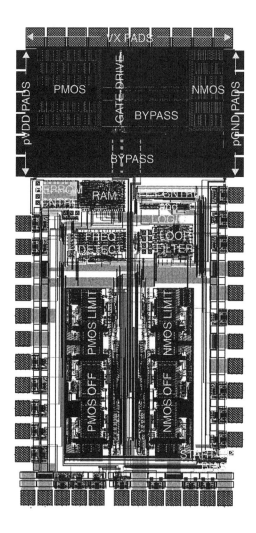

FIGURE 8.21 DVS Regulator chip layout.

Measured Results

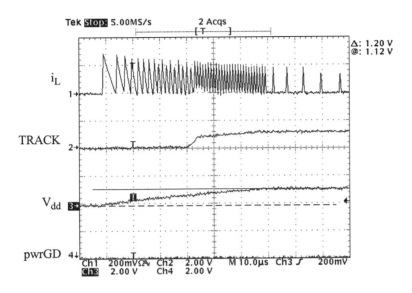

FIGURE 8.22 Start-up transient from $V_{dd} = 0$ to $V_{dd} = 1.2$ V.

sion. The feedback loop intentionally slows the latter part of the transition to a first order decay, eliminating the possibility of undershoot.

TABLE 8.3 Tracking performance summary.

Transition	Tracking Time	Tracking Energy[1]
f_{VCO} = 5 MHz to 100 MHz V_{dd} = 1.08 V to 3.78 V	23.5 µs	4.6 µJ
f_{VCO} = 100 MHz to 5 MHz V_{dd} = 3.78 V to 1.08 V	44.0 µs	
f_{VCO} = 20 MHz to 40 MHz V_{dd} = 1.39 V to 1.82 V	7.3 µs	0.2 µJ
f_{VCO} = 40 MHz to 20 MHz V_{dd} = 1.82 V to 1.39 V	9.9 µs	

DC-DC Converter IC for DVS

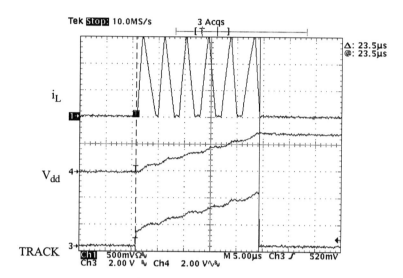

FIGURE 8.23 A 5 MHz to 100 MHz tracking transition.

TABLE 8.3 Tracking performance summary.

Transition	Tracking Time	Tracking Energy[1]
f_{VCO} = 40 MHz to 80 MHz V_{dd} = 1.82 V to 2.95 V	16.2 µs	1.2 µJ
f_{VCO} = 80 MHz to 40 MHz V_{dd} = 2.95 V to 1.82 V	19.3 µs	

1. Simulated for the full low-to-high-to-low tracking cycle

Table 8.3 summarizes the tracking performance for a variety of high-to-low and low-to-high frequency transitions at 1/4 full-load and V_{bat} = 6.0 V. Tracking time is measured from the rising to falling edges of the TRACK signal, yielding the 0% to $f_{des} - 3\text{ MHz}$ points. Tracking energy is for the entire low-to-high-to-low tracking cycle from C = 4.7 µF, the measured steady-state dissipation as a function of f_{VCO}, and the measured $V_{dd}(t)$ waveform.

8.7.2 Synchronous Rectifier Control

Figure 8.24 shows the i_L, v_{gn}, and v_x waveforms for a single PFM pulse at $V_{bat} = 3.3$ V, $f_{VCO} = 24$ MHz. The DC value of V_{dd} is 1.47 V. Here, the power NMOS is turned off at $i_L < 2$ mA, well within the error budget specified in Section 8.4.3, and introducing negligible LI_ε^2 loss.

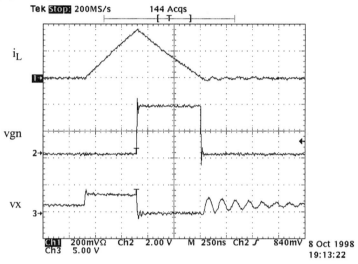

FIGURE 8.24 NMOS zero-current turn-off. $V_{bat} = 3.3$ V, $V_{dd} = 1.47$ V.

8.7.3 Regulation Performance

Figure 8.25 shows the regulation waveforms at $f_{VCO} = 24$ MHz, with $V_{bat} = 3.3$ V, under a large 22 mA load. The peak-to-peak output voltage ripple of 3.8% is near the anticipated value.

Figure 8.26 shows the measured full-load efficiencies for a variety of frequency requests, M.The numbers in this figure are generally consistent with expected results, though they tend to fall off at higher throughput requests. This is attributed to the higher-than-expected battery voltage, $V_{bat} = 5.0$ V (see Section 9.5), necessary to

DC-DC Converter IC for DVS

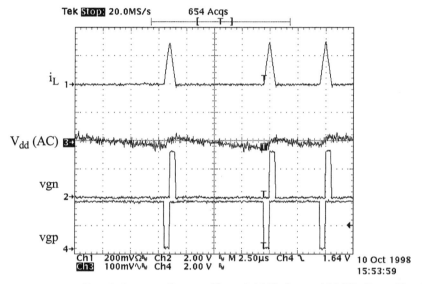

FIGURE 8.25 Regulation waveforms at $V_{dd} = 1.47$ V, $f_{VCO} = 25$ MHz, $I_{dd} = 22$ mA.

allow the 89 MHz and 100 MHz operating points (all other efficiency data was taken at V_{bat}=4.0V.), and to the additional series resistance of the 68LDCC package and through-hole socket. Figure 8.27 shows the mechanisms of power dissipation for various loads at $V_{bat} = 3.3$ V, $V_{dd} = 1.47$ V, and $f_{VCO} = 25$ MHz. The recorded efficiencies are 87%, 85%, and 74% for 22 mA, 11 mA, and 1 mA loads. Power train dissipation, which includes losses in the power FETs, package, and all external filter elements, dominates converter losses, even at light load. The VCO receiver and frequency detector are the largest contributors to controller dissipation at light load. Analog power, dominated by the NMOS off comparator, is the largest dissipater in the controller at heavy load. All power measurements correlate well with simulated results.

Measured Results

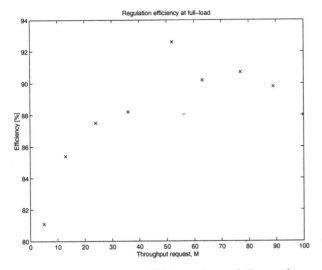

FIGURE 8.26 Efficiency in regulation mode.

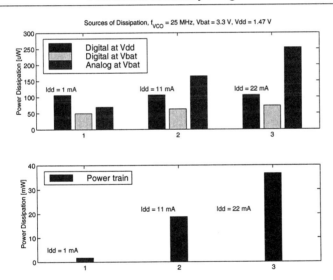

FIGURE 8.27 Mechanisms of dissipation versus load at 25 MHz and 1.47 V.

DC-DC Converter IC for DVS

References

[8.1] G. Wei, et. al., "A Variable-frequency Parallel I/O Interface with Adaptive Power Supply Regulation", *Proceedings of the IEEE International Solid-State Circuits Conference*, San Francisco, Feb. 2000, pp. 298-9.

[8.2] W. Namgoong, M. Yu, and T. Meng, "A High-Efficiency Variable-Voltage CMOS Dynamic dc-dc Switching Regulator", *IEEE International Solid-State Circuits Conference*, pages 380-381, Apr. 1997.

[8.3] T. Kuroda, et. al., "Variable Supply-Voltage Scheme for Low-Power High-Speed CMOS Digital Design", *IEEE Journal of Solid-State Circuits*, vol. 33, no. 3, pages 454-463, March, 1998.

[8.4] B. Acker, C. Sullivan, and S. Sanders, "Synchronous Rectification with Adaptive Timing Control", *Proc. IEEE Power Electronics Specialists Conference*, 1995.

CHAPTER 9 — *DVS System Design and Results*

A complete embedded microprocessor system was designed and implemented to demonstrate the processor system design methodology described in the previous chapters. By combining Dynamic Voltage Scaling with energy-efficient architecture and circuit design, the system is able to demonstrate more than an order of magnitude improvement in energy efficiency over more conventionally implemented designs.

In order to measure the energy efficiency of programs typically running on portable devices, a complete software infrastructure was developed. This infrastructure includes a pre-emptive multi-tasking real-time operating system providing standard C library functionality, which allowed standard C programs to be compiled for the system. A programmable I/O board enabled rapid prototyping of I/O devices to verify the system's functionality, and enabled multimedia programs with real-time constraints to run on the system.

Section 9.1-Section 9.3 describe the four chips as well as the physical board implementation. Section 9.4 describe the I/O board and the software infrastructure. Section 9.5 presents the measured performance of the prototype system, and Section 9.6 compares this system to other energy-efficient processors.

9.1 System Architecture

The prototype system, shown in Figure 9.1, contains four custom chips fabricated in a 0.6µm 3-metal CMOS process technology. The chip-set includes a microprocessor (Section 6.1), a battery-powered dynamically variable DC-DC voltage converter (Chapter 8), a bank of SRAMs (Section 6.3), and an interface chip for connecting to commercial peripheral devices, which are modeled by the I/O board. These chips integrate all the necessary logic for inter-chip communication so that they can be seamlessly connected together. For a completely functional processor system, the only external components required are a crystal oscillator, an inductor, and several small bypass capacitors.

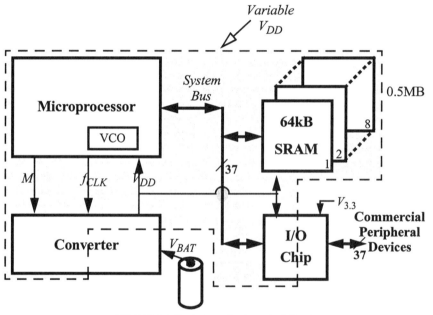

FIGURE 9.1 Prototype System Architecture.

As shown in Figure 8.1 the DVS voltage regulation loop consists of the battery-powered converter chip, and the VCO which is connected to the loop via the V_{DD} and f_{CLK} signals. The processor commands the desired clock frequency via the digital value, M. The 37-bit *System Bus* consists of a 32-bit multiplexed address/data bus, and five bits of control. In addition, the processor generates chip enable signals for the I/O and SRAM chips.

There are three voltage domains in the system. The converter outputs the variable DVS voltage, V_{DD}, which powers the processor, the SRAM chips, the I/O chip, and the front-end circuits on the converter chip. The battery voltage, V_{BAT}, supplies the converter's power FETs and back-end circuits. In addition there is a 3.3V voltage domain which supplies the output pads of the I/O chip, so that it can replicate the system bus at a standard voltage level, which allows the bus to connect to commercial IC's.

9.2 Interface IC

The primary function of this chip is to connect commercial, fixed-voltage peripheral chips to the variable-voltage system bus of the embedded DVS processor system. These chips may include ROM and DRAM, as well as chips providing system I/O, such as a serial communication controller (SCC), codecs, LCD controllers, etc. A StrongArm microprocessor and a Xilinx FPGA were used to model the I/O subsystem in the prototype system, and are described in more detail in Section 9.3.2. To simplify the design of interface chip, the bulk of the control FSM's to communicate with the StrongArm were pushed into the Xilinx connecting the interface chip to the StrongArm. Thus, the primary function of the interface chip is to level convert the system bus to a fixed 3.3V bus, and perform simple flow control. The level conversion occurs in the pads so that all the internal chip circuitry operates with the variable supply voltage, V_{DD}.

In a practical system implementation, this chip would be more complex in order to enable it to connect directly to peripheral I/O chips. With the controller circuitry integrated on-chip, the controller itself could be DVS compatible providing variable performance and energy consumption. Further enhancements would include having two regulator loops -- a processor core voltage/frequency, and an external memory system voltage/frequency. This would enable high-speed DMA transfers, when necessary, when the processor core is in a low-performance mode of operations.

To aid in system debugging, the processor system bus signals are always replicated on the 3.3V Xilinx-side bus. This allowed test equipment to monitor activity between the processor and main memory on the processor system bus, at a fixed voltage. In a practical system implementation, this feature would be optionally disabled in order to eliminate unnecessary energy consumption driving these signal pins when the I/O interface is not actively being used for either an I/O read/write or a DMA request.

The interface chip die, shown in Figure 9.2, measures 4.4 x 4.4mm, and contains 40k transistors, of which 5k are used by the controller implementation located in the center of the die. The chip is pad limited with its 132 I/O signals resulting in the large die size. The entire core outside the controller contains bypass capacitance used to bypass the two input voltage supplies (V_{DD} and $V_{3.3}$).

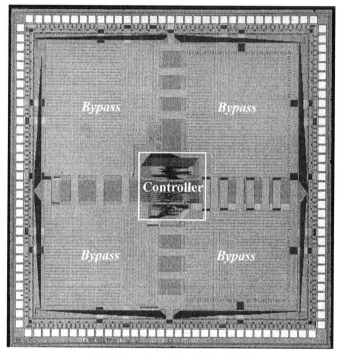

FIGURE 9.2 Interface Chip Die Photo.

The basic chip architecture is shown in Figure 9.3. When there is no active I/O or DMA request, the interface chip is in snoop mode. The bus clock (*MClk*), the processor bus (*PBus*), and the bus control signals (*Write, Byte, Burst, PWait*) all drive their equivalent Xilinx-side signals. All the on-chip signal paths are delay matched to maintain a constant delay shift across the Xilinx bus. To eliminate 15 unnecessary pins, the 16 memory chip enables (*CE[15:0]*) are OR-ed into a single signal, *XCE*.

For an I/O request, the processor asserts *IOCE*, which gets level-converted to *XIOCE*, and signals the Xilinx that an I/O request needs to be serviced. Because all the circuit paths forwarding signals from the processor bus to the Xilinx bus are delay matched, the Xilinx can interface to this bus in a synchronous manner since there is approximately zero relative delay shift on the Xilinx bus. This removed the need for an otherwise more costly asynchronous interface between the interface chip and Xilinx.

The timing for an I/O write is shown in Figure 9.4, which demonstrates how the inter-

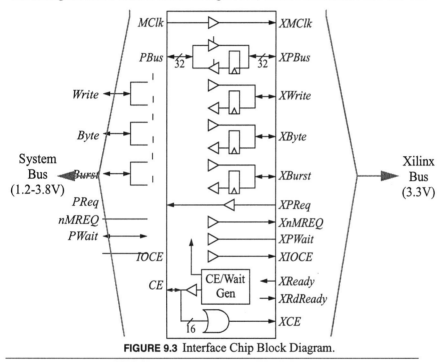

FIGURE 9.3 Interface Chip Block Diagram.

face chip interacts with the Xilinx chip. The transactions get replicated from the processor bus to the Xilinx bus, with the *XReady* signal providing flow control from the Xilinx back to the interface chip. By default, an I/O request will initially drive *PWait* high, stalling the processor system. Once the Xilinx has latched the address off of *XPBus*, it asserts the *XReady* signal one cycle, which in turn drives *PWait* low for one cycle, and advancing the state of the processor system one cycle. Once the data word has been transferred to the Xilinx chip, the transaction is complete. Because it can take many cycles to complete an I/O request, *PWait* is generally high for a majority of the duration of an I/O request.

DVS System Design and Results

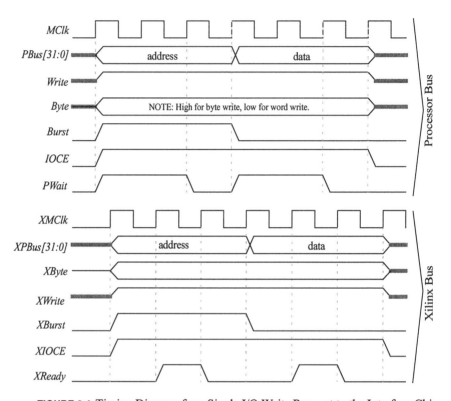

FIGURE 9.4 Timing Diagram for a Single I/O Write Request to the Interface Chip.

On an I/O read (Figure 9.5), the *XPBus* switches direction, as indicated when the *enXPBus* signal goes low, in order to receive the desired data. Since *XPBus* gets driven by the Xilinx delay-shifted with respect to *MClk*, it is latched and driven onto *PBus* the subsequent *MClk* rising edge, requiring the 32-bit latch to hold state for one cycle. To prevent both the interface chip and the Xilinx from driving *XPBus* at the same time, the *enXPBus* signal goes low a cycle early and stays low an extra cycle. As long as the delay through the interface chip is less than the cycle time of *MClk*, there will be guaranteed non-overlap times to eliminate this potential conflict on *XPBus*.

When the I/O subsystem wants to initiate a DMA request, the Xilinx asserts *XPReq*, which in turn asserts *PReq* and informs the processor of a pending DMA request.

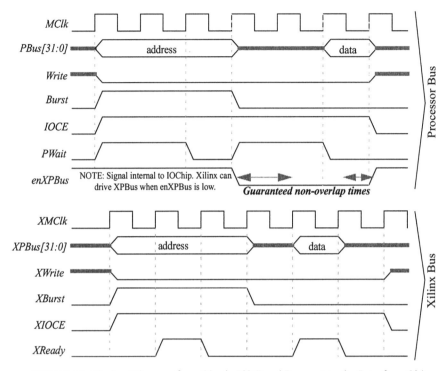

FIGURE 9.5 Timing Diagram for a Single I/O Read Request to the Interface Chip.

Once the processor has completed any outstanding bus access, it releases the processor bus and synchronously deasserts *nMREQ*, which then deasserts *XnMREQ* giving the I/O subsystem control of the bus. At the same time, the direction of the control signals is changed, and they are driven by the Xilinx via *XWrite*, *XByte*, and *XBurst*. Similar to *XPBus*, these are latched in order to resynchronize these signals with the edge of *MClk*. The Xilinx does not need to drive the *CE* signals, as they are internally generated by the interface chip, which can infer these signals by decoding the address placed on *XPBus*. In DMA mode, the *XRdReady* signal is used to indicate when the SRAM has returned the value of a DMA read request.

The interface chip was placed into a 132-pin QFP package. There are 103 signal pins, with 56 pins required for the processor system bus and 44 pins required for the Xilinx bus. An additional two pins are utilized for debugging, and one pin for the reset signal. The remaining 29 pins are used for ground and supply lines. The active circuit area on the interface chip was only approximately 2 mm². The large die size was necessary given the large number of pins required to interface between the two busses.

9.3 Prototype Board

The prototype system was constructed on an 8-layer 6" x 8" PCB board, with four supply layers, and four routing layers. Due to the integration of the memory and interrupt controllers onto the processor chip, few external components were required to construct the system. Extra complexity was added for features which supplemented system debugging, such as bypassing the converter chip with a fixed external voltage, and split core (V_{DD}) and I/O supplies (V_{DDIO}). The prototype system communicates to a StrongArm-based system board (Section 9.3.2), which emulates I/O activity, via a DB2x25 connector.

9.3.1 Test Board

The 4 unique custom IC's of the prototype system are connected as shown in Figure 9.6. The 37-bit system bus connects the processor chip to the SRAM chips and the interface chip. The nine chip enables (*CE[8:0]*, *IOCE*) are output by the integrated memory controller. An additional eight chip enables are available for SRAM chips, but were left unused in the prototype system. The interface chip communicates with the StrongArm board's Xilinx chip via a 43-bit bus, which replicates the system bus functionality with a few additional control signals. The system reset switch allows the processor system to be reset while leaving the converter actively operating, and can be used if the processor performs an illegal operation. The processor also uses 1 MHz oscillator to provide the reference frequency used by the internal real-time counter.

9.3.2 StrongArm I/O Board

The StrongArm board is a commercial development board, and used to model I/O from peripheral devices. A software approach was chosen so that I/O devices (e.g. codec, LCD, radio, etc.) could be rapidly constructed and modeled, as well as to provide low-level debugging functionality for the prototype system. The board allows all I/O output to be validated and time-stamped to ensure correct I/O output from the pro-

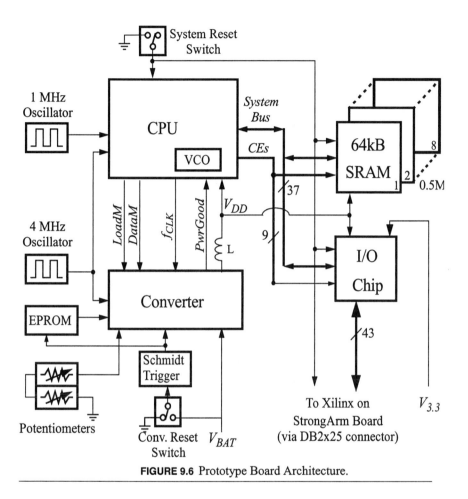

FIGURE 9.6 Prototype Board Architecture.

totype system. In addition, the StrongArm board can generate input data at set intervals, much like any I/O device would.

The ARM programming environment provides debug and monitoring support. The current version, Angel, is used by the StrongArm board, and its predecessor, Demon, is used by the prototype processor. This debug and monitoring support consists of low-level software running on the host CPU, and remote software running on a PC or Sun workstation. The software communicates via a serial channel, which is emulated for the prototype system by the StrongArm, and allows the debugger to properly operate on the prototype processor.

9.4 Software Infrastructure

To fully qualify the energy-efficiency improvement of DVS, a software environment typically found in a portable device was booted on the prototype system. This includes a real-time operating system (RTOS), the voltage scheduler required by DVS, and common application programs. The prototype system would then execute the benchmark application with and without the voltage scheduler to quantify the increase of processor system energy-efficiency due to DVS. There is a more in depth discussion of the software required for this system in Chapter 10.

9.4.1 Software stack

The software infrastructure stack is shown in Figure 9.7. On the prototype processor is the low-level Demon debug monitor, on top of which sits the RTOS, the voltage scheduler and the user application programs. On the SA-1100 is the Angel debug monitor, which sits underneath the StrongArm I/O Processor (SAIOP) software program. This provides I/O support to the RTOS, and creates the virtual channel which allows Demon to communicate with the remote debugger program, armsd, running on a Sun. Another armsd program running on a PC interacts with the SA-1100 debugger. The RTOS described in Section 10.5 is a custom pre-emptive multi-tasking kernel that contains a temporal scheduler and standard C library functionality. The temporal scheduler decides which task runs when using an earliest-deadline-first (EDF) policy, which is optimal for fixed speed systems (Section [9.2]). The kernel is not cognizant of the speed setting of the processor. Whenever the temporal scheduler updates the process schedule, the voltage scheduler is executed, which is run as a separate thread on top of the kernel. The voltage scheduler analyzes the current process schedule and application deadlines to provide a voltage schedule for varying microprocessor performance. The algorithm is discussed in further detail in Section 10.6.

The user applications are written in C/C++ using the full C library support provided by the RTOS. The three application used in the DVS evaluation benchmark suite (MPEG, UI, AUDIO) are discussed in Section 10.4.

9.4.2 Software I/O processor (SAIOP)

The RTOS and user applications use address mapping, as described in Table 9.1, to specify the destination for I/O data. The SAIOP program then routes the I/O data to the desired location on the SA-1100.

TABLE 9.1 IO Space Address Mapping.

I/O Device	Address	Description
IO Channel	0x48003a00	I/O control information. Opens/closes a file or network connection, and performs flow control
	0x48002000	I/O read data.
	0x48002400	I/O read control. Provides information on channel, and μs delay until next word is to be read.
	0x48003b00	I/O write data. Data word is tagged with channel being written to.
Frame Buffer	0x58xxxxxx	Frame buffer. Writes to this space are logged in framebuffer.dat file for later verification.
	0x78xxxxxx	Frame buffer color-map. Write to this space adjust the color-map of the display device.
Debug Space	0x680001xx	Debug space. Used for low-level debugging of RTOS state.
Serial Channel	0x880000xx	Serial channel. Mimics the register set and functionality of a standard UART serial interface.

With the exception of the serial channel and frame buffer, all I/O connections are established as sockets, and time-share a single I/O channel location. Each I/O device is allocated a unique channel ID, which is used to tag all input/output data on the I/O channel for that device. Flow control is available to slow down and/or speed up the flow of data as necessary. Writes to the I/O channel are verified against the master data set stored in the Flash ROM, and reads have their data supplied by the ROM, and tagged with a delay time for which the SAIOP should wait until asserting the interrupt line to indicate that the next data word is ready.

The frame buffer is located in a separate address space, and the contents thereof are written to a file for post-execution evaluation to ensure the correct data was written to

DVS System Design and Results

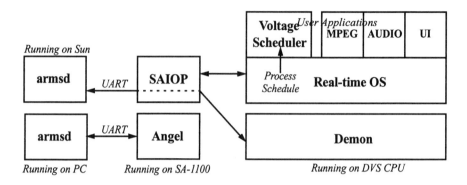

FIGURE 9.7 Software Architecture.

it. The debug space is used to perform low-level thread and speed tracing of the prototype processor, which aided in the debugging of the system. The SAIOP maps the virtual serial channel to the physical UART on the StrongArm board, allowing the Demon running on the prototype processor to communicate with the remote debugger on a Sun workstation.

9.5 Evaluation

The prototype system was used to provide a test bed to allow measurements on a hardware system to verify the simulations used in the design process. In addition, the benchmark program Dhrystone 2.1 was run in order to measure the energy consumption in terms of MIPS/Watt, a commonly quoted measure, to compare against commercial processor implementations.

While the original design target was for V_{DD} to operate over 1.1-3.3V with a clock frequency range of 5-100 MHz, the prototype silicon failed to operate for a V_{DD} less than 1.2V. Since even the VCO failed, which consists of only CMOS pass gates and inverters, the most likely cause of failure was a much larger $|V_{Tp}|$ than specified in the process manual (0.95V, worst case, with a 0.7-1.5V wafer acceptance range), although no test structures were on the die to verify this hypothesis. However, the prototype system successfully operated over the voltage range 1.2-3.8V, although over the somewhat lower frequency range of 5-80 MHz, as shown in Figure 9.8, demonstrating the ability of a DVS processor system to scale with widely-varying process parameters.

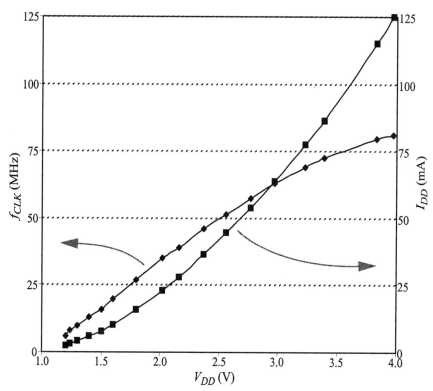

FIGURE 9.8 Measured Clock Frequency and Supply Current vs. Supply Voltage.

9.5.1 Transient operation

Figure 9.9 shows a scope trace for the system's maximum low-to-high and high-to-low speed transitions for the regulator described in Chapter 8. The V_{DD} signal transitions from 1.2V to 3.8V, then back down to 1.2V. The *Track* signal indicates whether the converter loop is in the tracking mode, in which it is actively changing V_{DD}, or in regulation mode, in which it is trying to maintain a constant V_{DD} value. This signal demonstrates that the maximum transition time is 70μs for the 5-80 MHz transition under full system load, while smaller voltage transitions can be performed in less time. During this entire transition period, the processor system can continue to execute instructions.

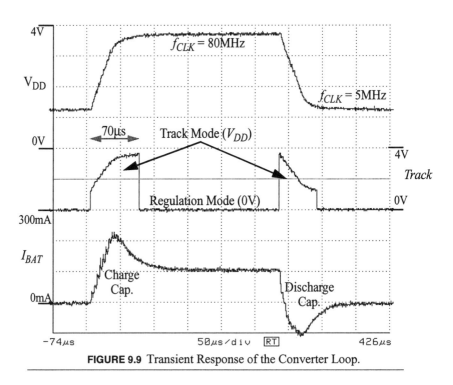

FIGURE 9.9 Transient Response of the Converter Loop.

The decaying exponential response of V_{DD} demonstrates that the converter loop behaves much like a single dominant-pole system. In fact, V_{DD} changes to within 70% of its final value within only 25µs, because it is slew-rate limited to 0.08 V/µs.

The signal I_{BAT} is the battery current measured going into the regulator, but after the battery's bypass capacitor. There is a current spike on the low-to-high transition which is required to charge up the loop's output capacitor to the required voltage. The negative current spike on the high-to-low transition occurs because the power PMOS is removing charge from the output capacitor and placing it back onto the battery's bypass capacitor at approximately 90% conversion efficiency (see Section 8.8). The conversion loss of the loop is the transition energy, which is a maximum of 4μJ for both the low-to-high and high-to-low transitions.

9.5.2 Dhrystone Benchmark

The Dhrystone 2.1 benchmark is commonly used for microprocessors in embedded applications to characterize throughput in MIPS, as well as energy consumption in Watts/MIP Section [9.3]. This benchmark was compiled for the prototype system, so that system energy-efficiency could be directly compared against the energy-efficiency of commercial ARM implementations.

Figure 9.10 plots the prototype system's throughput versus its energy consumption for the Dhrystone 2.1 benchmark. The upper curve is for the system when it is powered by a fixed, external voltage source, and the converter is disabled. The lower curve is for the system with the converter loop enabled. The curves are generated by running the system at constant frequency and V_{DD} to demonstrate the full operating range of the system. The throughput ranges from 6-85 Dhrystone 2.1 MIPS, and the total system energy consumption ranges from 0.54-5.6 mW/MIP. The efficiency of the converter loop, which is proportional to the gap between the two curves, ranges from 90% at high voltage to 80% at low voltage.

With DVS, peak throughput can be delivered upon demand. Thus, the true operating point for the system lies somewhere along the dotted line because 85 MIPS can always be delivered when required. When only a small fraction of the computation requires peak throughput, the processor system can deliver 85 MIPS while consuming, on average, as little as 0.54 mW/MIP.

A commonly quoted energy-efficiency metric is MIPS/Watt. The equivalent for this system would be the ratio of peak MIPS to average power dissipation because the throughput and power dissipation can be dynamically varied. In the optimal case when peak throughput is required only a small fraction of the time, the system's average power dissipation can be as low as 3.24mW, yielding 26,200 MIPS/W. When the system is operated at constant V_{DD}, the energy-efficiency is a maximum of 1,850 MIPS/W at 1.2V.

9.5.3 Idle Energy Consumption

Because a microprocessor in portable systems idles a significant amount of time, the energy consumed while idling can become critical to the overall energy efficiency.

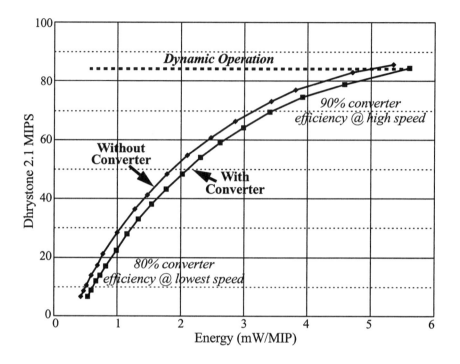

FIGURE 9.10 Measured Throughput vs. Energy Consumption.

For the prototype system, a halt instruction was implemented via a coprocessor write instruction, which asserts the *Sleep* signal. This signal effectively stops all activity by clock gating the rest of the system, with the exception of a few state registers in the interrupt controller, the external bus interface, and the real-time counters.

If the processor speed is set to 5 MHz before entering sleep, the entire system dissipates only 800µW of power, with a one cycle start-up from sleep. The latency to ramp back up to full speed upon wake-up is set by the converter loop to be 70µs, although the processor can continue operating during this ramp up period and begin immediate execution of the interrupt handler.

Evaluation

9.5.4 DVS benchmarks

To evaluate DVS, benchmark programs were chosen that represented software applications that are typically run on notebook computers or PDAs. Existing benchmarks (e.g. SPEC, Dhrystone MIPS, etc.) are not useful because they only measure the peak performance of the processor. New benchmarks were selected which combine computational requirements with realistic latency constraints. The three programs are MPEG, UI, and AUDIO, and are described in more depth in Section 10.4.

Measuring Energy Consumption

To measure energy consumption of the benchmark applications, the simple circuit in Figure 9.11 was used in-line on the regulator's voltage supply, V_{BAT}. After Demon boots and the RTOS and benchmark program are downloaded into main memory, the Demon break-points the start of the application and idles at low voltage. When instructed by a "go" command from armsd, the benchmark will execute, and at the end of running, will put the processor back into idle mode at low voltage.

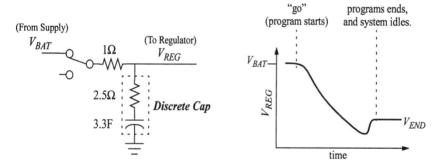

FIGURE 9.11 Energy Measurement Circuit & Transient Response.

While Demon is booting, the switch remains closed and the capacitor maintains V_{BAT} across its terminals. At the break-point, the switch is opened, the "go" command is given, and V_{REG} roughly changes as depicted. The voltage drops due to the capacitor sourcing charge, and due to an IR drop on the intrinsic resistance of the discrete capacitor. When the microprocessor idles after completing the application, V_{REG} jumps back up a little bit due to the IR drop disappearing and settles to V_{END}. During low-voltage idle, the drop on V_{REG} is 60 μV/sec, and hence, very flat.

DVS System Design and Results

The energy consumption of the benchmark is:

$$\Delta E = \frac{1}{2} \cdot 3.3F \cdot (V_{BAT}^2 - V_{END}^2) \qquad \text{(EQ 9.23)}$$

There is energy loss in the 2.5Ω resistor, but at the maximum average current of 20mA, the loss is only 1.2%, which was neglected. Thus, the energy consumption of the benchmarks could be measured to within 99% accuracy.

Results

Using the above approach for measuring energy consumption, the three benchmarks were first run at constant maximum throughput to measure the baseline energy consumption. They were then re-run with the voltage scheduler enabled, and had their energy consumption measured again.

TABLE 9.2 Measured Benchmark Energy Consumption (Normalized).

	Benchmark Programs		
Algorithm	MPEG	UI	AUDIO
Maximum Performance	100%	100%	100%
Optimal	67%	25%	16%
Voltage Scheduler	89%	30%	22%

Table 9.2 shows the measured system energy consumption for the three benchmarks, which is normalized to when the system is running at maximum throughput, since this is the typical operating mode of a processor system that operates from a fixed V_{DD}. The row labelled Optimal is the energy reduction when all the computational requirements are known a priori, and is an estimated value derived from simulation. The optimal values represent the maximum achievable energy reduction for these benchmarks. The last row is the measured energy consumption with the voltage scheduler enabled. As expected, the compute-intensive MPEG benchmark has only a 11% energy reduction from DVS. However, DVS demonstrates significant improvement for the less compute-intensive AUDIO and UI benchmarks, which have a 4.5x and 3.5x energy reduction, respectively. Comparing the DVS results against the optimal results demonstrates that while the voltage scheduler's heuristic algorithm has a difficult time optimizing for compute-intensive code, it performs extremely well on non-speed critical applications.

Table 9.3 shows the average power dissipation of the three benchmarks with the voltage scheduler operating. The effective MIPS/W is calculated as the ratio of peak throughput (85 MIPS) to average power dissipation, and demonstrates the achievable increase in energy efficiency when the system is running real programs. Both the UI and AUDIO benchmarks have an average power dissipation on the order of 10mW, yielding an energy efficiency on the order of 10,000 MIPS/W.

TABLE 9.3 Measured Power Dissipation with the Voltage Scheduler.

	Benchmark Programs		
Voltage Scheduler:	**MPEG**	**UI**	**AUDIO**
Average Power (mW)	145	11.75	8.00
Effective MIPS/W	600	7,200	10,600

Thus, real applications, with the proper operating system support via the voltage scheduler, can achieve a significant reduction in energy consumption with DVS, thereby improving processor system energy-efficiency by up to a factor of 10x.

9.6 Comparisons and other related work

A technique for minimizing the supply-voltage to reduce energy consumption utilizing a voltage regulator was initially proposed for digital circuits at fixed throughput [9.4]. A replica of the critical path was used in a negative-feedback loop to set V_{DD} to the lowest possible level, while the circuits continued operating correctly given a desired clock frequency.

This technique was subsequently demonstrated on a MIPS R3900 processor core, with an integrated, on-chip, voltage regulator [9.5]. A desired operating frequency is set externally, and the regulator outputs the minimum V_{DD} value at which the processor core can continue operating. However, the clock frequency could only be set externally, and requires a system reboot in order to change the frequency value.

This technique was enhanced to dynamically scale V_{DD} for variable-rate digital signal processing [9.6]. A variable-rate processing circuit has an input FIFO, which is monitored for how full it is. When the FIFO is near empty, V_{DD} can be reduced, and when the FIFO is near full, V_{DD} must be increased to catch up to the input data. Adaptive scaling was later demonstrated with an open-loop regulation approach, which used

DVS System Design and Results

four V_{DD} values to provide faster switching transients and used dithering to approximate intermediate V_{DD} values *[9.7]*. More recently, an approach for dynamic voltage scaling has been demonstrated for I/O interfaces, in which V_{DD}, and the energy consumption, scales with the throughput demands on an I/O transceiver *[9.8]*.

Comparison to Other ARM Processors

Another goal of this work was to see how much the intrinsic energy-efficiency of a microprocessor could be improved without the benefit of DVS. A key benefit of implementing a commercial ARM8 processor core was that the prototype processor could be compared against other commercial ARM microprocessor implementations. One of these implementation is the StrongArm SA-110, which is the most energy-efficient commercial microprocessor available to date.

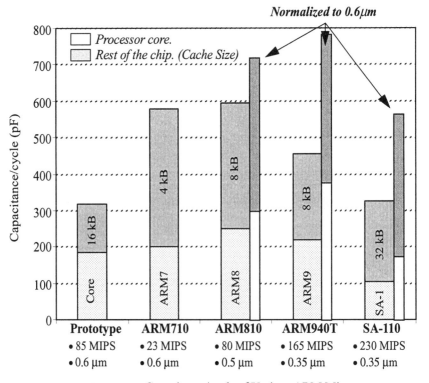

FIGURE 9.12 Capacitance/cycle of Various ARM Microprocessors.

302   *Energy Efficient Microprocessor Design*

Energy consumption, in capacitance/cycle to normalize out the dependence on V_{DD}, is plotted for the prototype processor and four commercial ARM microprocessors in Figure 9.12. In addition, the processors performance and process technology is given. The capacitance/cycle is broken out to show that which is consumed by the core, and that which is consumed by the cache and the rest of the chip.

The prototype processor has the lowest capacitance/cycle of the five implementations, with the SA-110 a close second. However, since the ARM810, ARM940T, and SA-110 have the benefit of a better CMOS process technology, the capacitance/cycle for these three were normalized to the 0.6μm process technology of the prototype processor. Compared against the normalized values, the prototype processor demonstrates almost 2x lower capacitance/cycle than any of the four commercial processors, validating the energy-efficient design methodology presented in this work. Despite the microarchitectural constraint of using the ARM8 processor core, the prototype system was still able to demonstrate a significant reduction in capacitance/cycle.

Since the cache sub-system was designed in its entirety with energy-efficiency in mind, it is interesting to see how the non-core component of the prototype processor's capacitance/cycle compares against the other implementations. Comparing the normalized values, the non-core component of the prototype processor is 3x lower than any of the other commercial implementations, despite having a larger cache size than all but one of the other implementations. Thus, if the processor core itself was re-architected and the instruction-set architecture (ISA) designed with energy-efficiency in mind, an even more energy-efficient microprocessor could be achieved.

References

[9.1]   Hewlett Packard, *CMOS 14TA/B Reference Manual*, Document Number #A-5960-7127-3, Jan. 1995.

[9.2]   C. Liu and J. Layland, "Scheduling Algorithms for Multi-Programming in a Hard Real-time Environment", *Proceedings of CACM 20*, 1973.

[9.3]   R. Weicker, "Dhrystone: A Synthetic Systems Programming Benchmark", *Communications of the ACM*, Vol. 27, No. 10, Oct. 1984, pp. 1013-30

[9.4]   V. von Kaenal, P. Macken, and M. Degrauwe, "A Voltage Reduction Technique for Battery-operated Systems", *IEEE Journal of Solid State Circuits*, Vol. 25, No. 10, Oct. 1990, pp. 1136-40.

[9.5] T. Kuroda, et. al., "Variable Supply-voltage Scheme for Low-power High-speed CMOS Digital Design", *IEEE Journal of Solid State Circuits*, Vol. 33, No. 3, Mar. 1998, pp. 454-62.

[9.6] L. Nielsen, C. Niessen, J. Sparso, and K. van Berkel, "Low-power Operation Using Self-timed Circuits and Adaptive Scaling of the Supply Voltage", *IEEE Transactions on VLSI Systems*, Vol. 2, No. 4, Dec. 1994.

[9.7] A. Chandrakasan, V. Gutnik, and T. Xanthopoulos, "Data Driven Signal Processing: An Approach for Energy Efficient Computing", *Proceedings of the 1996 International Workshop on Low-Power Design*, Aug. 1996, pp. 347-52.

[9.8] G. Wei, et. al., "A Variable-frequency Parallel I/O Interface with Adaptive Power Supply Regulation", *Proceedings of the IEEE International Solid-State Circuits Conference*, San Francisco, Feb. 2000, pp. 298-9.

CHAPTER 10

Software and Operating System Support

10.1 Software Energy Reduction

The basic goal of energy reduction from the software standpoint is to maximize the battery lifetime of portable general-purpose microprocessor devices by reducing the energy necessary to complete a given task without significantly changing system behavior. Software energy reduction techniques can be divided into two categories: static, which optimizes software before it is executed, and dynamic, which alters the operation of the device at run-time. Static energy reduction techniques for a microprocessor in a general-purpose system can be divided into two categories: high-level application design and compile-time optimization. Dynamic techniques rely on the software running in a portable electronic device to monitor and adjust the device operation at run-time. For example, a laptop that can turn off its LCD backlight requires a dynamic algorithm to determine when the backlight is not needed. Typically, dynamic techniques also require some modification to the base hardware, i.e. the ability to turn off the LCD display, which might not inherently reduce energy consumption. DVS is a dynamic technique that controls the speed of the CPU.

Application-level static techniques effectively trade-off the software's flexibility and functionality: a very simple application, although extremely energy efficient, will not provide the complex functionality required by some users. Similarly, advanced programming techniques, such as high-level object-oriented programming languages, may not be as efficient as simpler techniques but enable a programmer to quickly cre-

Software and Operating System Support

ate flexible programs. There is no quantitative answer to the flexible vs. efficient design decision: the benefit of advanced applications must be weighed against the annoyance of reduced battery lifetime. However, programmers which are aware of the trade-off will be able to design better applications by simply not including expensive features with marginal benefit.

At the highest level of application-level energy optimization, an application can be statically partitioned so that only a portion of it executes on the portable device, relegating the remaining processing to a wired compute server with access to a plentiful energy source and uses a wireless communication channel to connect the two halves. For example, the InfoPad Project *[10.1]* treats the battery-operated portable component as a mere display device: the application itself runs on a compute server and sends graphical screen updates across a wireless link to the portable device. Since the terminal only requires the ability to display transmitted bitmap data, not the resources to execute the entire application, it can be highly optimized and designed to minimize energy consumption. Moving the core application and optimizing the portable hardware allows the InfoPad portable terminal to provide full functionality without consuming large amounts of power.

Compile-time energy optimization entails reducing the energy consumed by choosing the optimal low-level machine representation for a given piece of high-level program. Assuming a simple machine model, which is adequate to describe most processors found in battery-operated devices, the easiest way to accomplish this is to minimize the number of instructions executed while maintaining correct program behavior: each instruction requires a significant amount of energy to fetch and decode, which is relatively independent of what is being executed. Traditional speed-based compiler optimizers are good at reducing the number of instructions; therefore, we can conclude that further energy-aware compiler optimizations will be very difficult, although possible because the energy consumed by each executed instruction is not exactly constant.

Turning a portion of the system completely on and off is the simplest form of a dynamic energy reduction technique: turn off the disk drive when not in use, turn off the display backlight when not in use, gate the clock to the microprocessor when it is idle, etc. The challenge of these dynamic approaches is to determine when to turn the device off and when to turn it back on again. For example, since a user may still be looking at the computer display after they stop typing, a heuristic must be used to control the device: turn off the display if the user has not typed anything for fifteen minutes. Microprocessors typically present several different levels of "off," which each has a different wake-up time and consumes different amounts of power. Choosing between these levels requires a similar heuristic: gate the processor clock when there are no useful instructions to execute and put the processor into a deep-sleep mode after it has been idle for twenty seconds.

Exploiting the continuous range between the discrete "on" and "off" states, by providing different *rates* of execution, is the next logical step for dynamic energy reduction in a microprocessor system (*[10.2]*). For example: run the processor as fast as possible when there is a lot of work to be done, but only run it at 25% speed when there are just a few background maintenance tasks to perform. By reducing the processor operating voltage along with its speed, significant energy savings can be realized however this requires an enhancement to the operating system which analyzes the current, past, and predicted computation requirements to determine the voltage at which to run the processor and is called the voltage scheduler*[10.3]*.

A complete DVS microprocessor system, consists of a DVS-capable microprocessor and related operating system software. Although the hardware and software designs are largely independent, developed using entirely different source languages and specifications, the entire flow to reach an efficient and effective final implementation requires considerable collaboration between the two components. For example, the processor can be optimized using "real" applications, instead of contrived test cases, and the analysis of voltage scheduling algorithms using an accurate hardware simulation environment allows optimization before a working physical platform is realized. The final integration stage requires the integration of the complete hardware system, including I/O functionality, with the desired applications and support software.The target hardware used for evaluation is the DVS prototype described in Chapter 6 and Chapter 9.

The software environment consists of an embedded real-time operating system, application software, support libraries, and an instruction-level microprocessor simulator. The benchmark application suite was formed based loosely on the requirements of a portable wireless terminal that supports multi-media access and simple user-interface applications. Applications were initially developed native on a high-performance workstation and then ported to the simulation environment. The environment and application code was based on pre-existing publicly available source code as much as possible to minimize the development effort.

The simulator is based on a standard ARM simulator extended with voltage-aware energy estimation and a memory-I/O subsystem that accurately emulates the hardware design. Energy estimation was used to optimize the hardware design, by recording the usage statistics of different functional units, and to evaluate different scheduling algorithms by providing application energy consumption estimates based on run-time simulations. The design of the memory subsystem, which consumes a significant portion of the total energy, is critical: using an accurate cache model allows various trade-offs, such as line-size or associativity, to be easily explored. The I/O subsystem is important because it affects application timing, e.g. waiting for wireless communication or user input, which alters the behavior of a voltage scheduling algorithm.

10.1.1 Software for Dynamic Voltage Scaling

The energy, E, consumed by a CMOS microprocessors can be calculated using the equation $E \propto N_{ops}CV^2$, where N_{ops} is the number of operations executed, C is the switched capacitance per operation, and V is the operating supply voltage (see Chapter 2). Architectural energy-reduction techniques, such as clock-gating, discussed in Section 4.1 focus on reducing C, while traditional compiler optimizations minimize N_{ops}. DVS, focuses on reducing V, by reducing the operating voltage to the maximum extent possible consistent with the application requirements.

The *voltage scheduler* is the part of the operating system that is responsible for analyzing the processing requirements of individual threads, overall system loading, and resource utilization in order to determine the speed at which a DVS processor should run. There are two basic varieties of voltage scheduler analyzed in this work: interval-based, which operate by examining the high-level system utilization, and thread-based, which operate using the requirements of each individual thread in the system. Thread-based schedulers tend to perform better for complex or varied systems, but require greater implementation and run-time overhead.

To determine the ideal processor speed, a scheduler needs to know the processing requirements of a task, measured in cycles, and its associated deadline, which is by when the processing needs to be completed. Ideally, a voltage scheduler would know the exact requirements and deadlines for all running tasks in advance, allowing it to compute an optimal schedule; however, this information is typically not available in a real general-purpose system. Voltage schedulers, therefore, must use estimates or sampling techniques to heuristically calculate the processor speed.

For example, if no task-level information is available, then the scheduler must either assume the worst and run the processor at full speed, or use a time-based sampling technique to gauge the processor load. Furthermore, if accurate information about task characteristics, for example if they are periodic or sporadic, is unavailable, then the scheduler must make an assumption that will likely cause it to over- or under- estimate the processing required. Our results show that it is possible to effectively schedule a system, and realize significant energy savings, without complete task-level knowledge.

A complete system implementation is necessary to fully understand DVS since it alters both hardware performance and software structure. For example, if one did not implement real benchmark applications, one would not accurately capture realistic frame-to-frame execution variances; if one did not incorporate a voltage scheduler implementation with a complete operating system, one could not measure the introduced increase in scheduling overhead; and if one did not encompass low-level hardware interactions, one would not include I/O interaction which fundamentally behaves differently than processor-bound execution and could skew results.

Three benchmark programs and four scheduling algorithms are used to characterize DVS. The benchmarks applications, which we feel would be typical of a battery-powered portable electronic device, present a diverse workload, ranging from easy-to-predict multi-media to a sporadic user-interface, which helps to understand the behavior of the voltage schedulers. The benchmarks are based on publicly available libraries wrapped in custom application shells. Some of the voltage scheduling are based on previous work and one has been developed by the authors. These algorithms assume different levels of information presented by applications, and make different assumptions about application behavior, yielding results that are fairly similar or dramatically different, depending on specific application constraints. Output traces from the simulator are analyzed in detail to understand the algorithm's behavior; the conclusion suggests some directions for future research based on these analyses.

10.2 Software Environment

The software environment for the target device consists of a embedded operating system, support libraries, and application code. An overview of the different components is presented here; the operating system and applications are discussed further in separate chapters. The operating system is the only component significantly effected by DVS, although some application support is necessary for effective thread-based voltage scheduling.

10.2.1 Operating System

The run-time operating system for the DVS system is based on the standard run-time ARM library distribution, *armlib*, extended with multi-threading and real-time capabilities. The real-time environment is soft real-time (it does not provide any strict execution time guarantees), based around a preemptive deadline/priority hybrid scheduler. Basic ANSI-C primitives, such as memory allocation, basic library calls, and formatted text input/output are supported by the library. Voltage schedulers are implemented as independent threads which require the ability to examine the state of all other threads in the system and change the operating speed of the processor. Section 10.5 presents the operating system infrastructure and functionality in detail.

10.2.2 Support Libraries

Several standard libraries support application development in the software environment. The source for these libraries, whenever possible, was obtained from publicly available sources to minimize the overall development time and minimize licensing related distribution restrictions.

- **ssLeay**: cryptography library supporting IDEA, DES, and RSA encryption/decryption *[10.6]*.
- **gdip**: graphics toolkit based on a gif-manipulation library, *gd [10.9]*, and modified to support multi-threading and to work with the LCD frame-buffer. The library enables simple graphics operations such as line- and text- drawing.
- **ipWin**: windowing environment developed in-house which supports simple multiple-access to the LCD display. Ostensibly, the interface is similar to X-Windows, presenting an event-based model with identical function names, but only simple operations are supported (i.e. no graphics clipping or fancy line styles).
- **ipTk**: simple in-house graphical user interface (GUI) toolkit presenting buttons, edit text, and labels, which provide the minimal functionality necessary for simple interactive applications. Using a pre-existing toolkit, such as tcl/TK or the Athena X-Windows toolkit, was not feasible because it would require functionality no supported by the run-time operating- or windowing- system.
- **tcl**: a publicly available *[10.14]* shell-like interpreter suitable for writing small scripts or simple applications. The tcl version used was not thread-safe, making it unsuitable for writing benchmark applications; however, execution of the tcl interpreter was used to profile execution to help optimize the processor.

Software Environment

- **ipSys**: library providing basic system functionality, such as interaction with I/O subsystems, which is not fundamentally a part of the core operating system. Most notably, it contains a standard FIFO implementation which is used extensively throughout the system to provide inter-thread communication.

10.2.3 Application Code

The applications are used to benchmark DVS algorithms as well as to help optimize the processor design. These two uses are sensitive to different execution characteristics; it is important that benchmarks adequately exercise the system in order to enable accurate analysis. DVS benchmarking, which is very sensitive to the timing of applications, what processing happens when, requires applications with complete user-level interaction semantics, such as modeling frame delays or the timing of user input events. Optimizing the processor design, however, is sensitive to the content of the applications, the specific instructions executed and cache access patterns, which is effected by the completeness of a benchmark program.

Three main benchmark applications were used in the design of the DVS prototype hardware and scheduling algorithms. These applications, covered in more detail in Section 10.4, form a diverse set of programs that exhibit varying timing and content characteristics.

- **Audio Decryption**: a stream of encrypted audio data is received over the wireless link, buffered, decrypted, and then transferred to the audio output device. The complete benchmark used is the IDEA decryption of a 10-second 11 KHz mono audio stream divided into 1 kB frames, yielding a 93 ms frame rate.
- **MPEG Playback**: an MPEG-2 movie received over the wireless link, buffered, decoded, and displayed on the LCD. The benchmark movie used, the beginning of a Simpson's Halloween Special, consists of eighty 192x144 frames displayed at 8 frame/sec.
- **GUI** - A simple address-book user interface allowing simple searching, selection, and editing operations. 432 user-interface events, such as pen-down or button press, are processed.

Software and Operating System Support

10.3 System Architecture

The target architecture consists of the DVS protoype processor (see Chapter 6), which implements the standard ARM instruction set, and a specific set of I/O devices, which are modelled around those that would be found on a typical portable electronic device. The exact I/O modules used are not critical to the evaluation of DVS: different systems could posses various I/O capabilities to support their individual applications. However, the choice of a specific hardware platform enables us to fully investigate the nuances of DVS by allowing us to create a complete implementation instead of performing an abstract analysis.

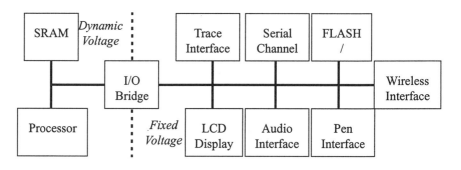

FIGURE 10.13 Hardware architecture.

10.3.1 I/O Subsystems

The hardware architecture of the target portable device, shown in Figure 10.13, provides several I/O components connected to the processor subsystem through a dedicated I/O bridge. The processor and SRAM operate under DVS, while the I/O subsystems operate at a fixed voltage and clock-speed. For most applications, the volume of I/O traffic is insignificant compared to the amount of computation performed (in terms of both energy and time); therefore, the performance of the I/O system, as long as it behaves reasonably, is not critical for an accurate analysis of DVS behavior. However, the functionality provided by the I/O system is critical as it strongly effects application execution.

System Architecture

- **Serial Connection**: This device provides debugging support by connecting the target platform to a host computer, allowing software to be downloaded and interactively debugged. This port is not intended to be used during normal operation, as it is not wireless and provides relatively low bandwidth.
- **Wireless Link**: The wireless link is the primary data source during operation. Source data for applications, such as Audio playback or MPEG decoding, must be transmitted across this link as there is no significant local data store. An approximately 1 Mb/s effective data rate is supported, which is comparable to currently available commercial systems.
- **Color LCD**: Direct-mapped color frame buffer and color pallet, supporting up to 640x480 with 8-bits/pixel. This device is used as a primary application data sink.
- **Pen-Input**: Pen-input packets are based on the output from industry-standard pen devices and are composed of five bytes which encode all necessary location and button-state information.
- **Audio Device**: Similar to the Color LCD, audio is used as a primary application data sink. It supports low-quality mono audio at 11 Khz with 8-bits/sample.
- **Trace Control**: The Trace Control I/O module is intended for emulation debugging and verification purposes only and will have no manifestation on a physical device; traces are saved by writing data through the simulator directly to disk. Several different levels of tracing are available, from cycle-level processor state, which writes the contents of the register file each cycle, to higher-level thread semantics, which registers an event each context switch.
- **EPROM/FLASH**: The EPROM or FLASH component provides storage for persistent content, such as programs or configuration data. The entire run-time software infrastructure requires under 2 MB of storage.

10.3.2 Processor Model

The processor simulator, instrumented to provide an "as accurate as possible" model for the hardware implementation, was used to evaluate performance across several architectural configurations and to estimate the energy consumption, both to optimize the processor as described in Section 5.2 and to develop and characterize the voltage scaling algorithms.

The prototype is based on the ARM8 processor architecture, which is an in-order single issue machine with a single instruction/data access path, static branch predictor, prefetch unit, and the ability to fetch two instruction words each cycle. The basic structure of the simulator is a giant case statement which decodes and processes each instruction type. In every cycle exactly one call to the memory model is made: an instruction fetch of either one or two words, a data load/store, or a null request. Except for incorporating energy estimation, discussed below, the simulator core was not changed.

10.3.3 Memory Model

The cache architecture used for the prototype processor is a unified 16 kB 32-way associative cache with 32-byte lines using a round-robin replacement policy and incorporating a 12-word write buffer. The design of the cache system, which has a major impact on system energy consumption, is discussed more thoroughly in Section 3.3. There is no programmable MMU, which would support access control on a page by page basis, included in the design. Memory accesses control, such as marking individual read/writes as non-catchable, is accomplished by setting the appropriate high-order address bits, providing the necessary functionality without significantly increasing hardware complexity.

The processor core simulator measures time strictly in terms of cycles, with exactly one memory system call each cycle. Therefore, in order to accurately model dynamic cycle-times caused by varying processor speeds, the accounting of real time (measured seconds) is handled in the memory model using a per-cycle time increment based on the current processor speed, dynamically set by the simulated voltage scheduling software.

10.4 Benchmarking

The system was evaluated using a customized benchmark suite that accurately models I/O interactions and exercises the complete software infrastructure. A detailed analysis of the relevant benchmark characteristics, such as cache miss rates and frame completion times, helps explain the behavior of different voltage scheduling algorithms. There are three aspects of the evaluation benchmarks, described below, that directly effect the operation of the processor: static characteristics, frame variance, and software overhead. The static characteristics are measurements which are independent of voltage scaling; frame variance is unique to DVS and directly effects its ability to predict the system workload; software overhead shows how applications interact with the underlying operating system. All the results in this chapter are generated using the cycle-level simulater discussed in Figure 5.2.

Two common multimedia tasks and a graphical user interface, which are all implemented on top of the OS, are used to evaluate the system:

- **Audio** - IDEA decryption of a 10-second 11 KHz mono audio stream.
- **MPEG** - Decoding of an 80-frame 192x144 MPEG-2 video at 8 frame/sec.
- **GUI** - A simple address-book graphical user interface application.

10.4.1 Static Characteristics

Table 10.1 summarizes the three benchmarks using several metrics that characterize the behavior of the application as a whole. These results are generated with the processor running at full speed, without voltage scheduling, to highlight properties that are inherent to the benchmark and not an effect of DVS. The "Cache Miss Rate" and "Core Stalled" parameters are primarily effected by memory system performance; therefore, they are instrumental indicators for designing an effective cache system. "System Idle" indicates the time that the system is not active, which relates to the energy consumed.

Several global statistics are useful for simply understanding an application. All values are measured with the processor running at full speed (100 MHz). The duration of the Audio and MPEG benchmarks is similar by design, enabling co-execution ("MPEG & Audio"), while the GUI takes much longer and must be executed separately.

Idle time indicates the potential for a reduction in clock speed: increased idle means that a voltage scheduler might be able to reduce the processing speed and stretch the computation to fill up the idle time. Since benchmarks with greater idle times also consume less energy because they execute fewer instructions during a given time interval; one might mistakenly assume that idle time is a good indication of an efficient system. However, idle time is really an indication that there is *opportunity* for energy savings through a reduction in the processor speed and voltage. Therefore, a

TABLE 10.1 Benchmark Characteristics

	Audio	MPEG	GUI
Cache Miss Rate	0.48%	0.66%	0.27%
System Idle	49.3%	12.1%	78.6%
Core Stalled (when system active)	7.7%	14.4%	12.2%
Energy Consumed	0.157 J	0.635 J	0.336 J
Duration	11.2 s	10.3 s	1769 s

Software and Operating System Support

system with large amounts of idle is executing inefficiently, and is a strong candidate for voltage reduction. Conversely, applications which have low idle percentages and consume the most energy are not greatly helped by DVS because their operating speed can not be reduced. Additionally, there are application-specific constraints which can also restrict the reduction of idle time, and so it is not a complete indicator of energy reduction potential.

10.4.2 Frame Characterization

The core work for each benchmark is divided into application-specific frames, which correspond to logical pieces of computation, to aid in both execution and evaluation. MPEG, for example, is inherently composed of video frames that naturally translate to frames of computation. Frames encapsulate and regulate computation: a deadline is determined for each frame, and the system attempts to complete frames by their associated deadlines by adjusting the processor speed (voltage scheduling) as well as the order in which threads are processed (temporal scheduling). The 'performance' of a given system can be measured in terms of how many frames miss their deadline, which will result in undesirable application performance, such as jerky video playback. In general, there is no advantage to completing a frame before its deadline (and it can also be a disadvantage, as discussed below). Each different application has an independent interpretation of a frame and associated missed-deadline penalty:

- **Audio** - The continuous data-stream is broken up into 1 kB frames, each representing 93 ms worth of audio. The size of each audio frame strikes a balance between reducing overhead, by using larger frame sizes, and reducing buffering requirements. A missed audio frame deadline results in skips, pops, or clicks: extremely noticeable and undesirable.
- **MPEG** - Composed of a stream of video frames, presented at 8 frame/sec. (125 ms deadlines). The frame-rate used is determined when the video is encoded: faster rates provide better quality at the cost of increased transmission bandwidth and processing requirements. A missed MPEG frame would result in jerky video playback, which is not quite as noticeable as missed audio frames but still undesirable.
- **GUI** - The user-triggered interface event, such as pen-down or key-press, naturally translates to a frame of computation, which starts with the event and ends when the display screen has been updated. GUI deadlines are determined by the human visual perception time: screen updates that occur 'too fast' are just as good as those that occur 'fast enough' and the user will not detect the difference. A deadline of 50 ms is used, loosely based on interface studies *[10.7][10.11][10.13]*, although the exact optimal value will vary from user to user. Missing a GUI frame

deadline results in a flicker on the screen, which can be anything from mildly to extremely annoying to the user, depending on their disposition and the exact delay incurred. Most GUI events, such as simple button updates, can easily complete before this deadline; some events, such as opening a new dialog window, are extremely computationally expensive and will always take a noticeable amount of time to complete.

The execution time of each frame within an application is not constant; the frame-to-frame variance can be analyzed to help understand how an application will react to voltage scheduling. Figure 10.14 diagrams frame completion histograms for each benchmark, further highlighting the inherent differences in their structure. Completion times are normalized to each benchmark's deadline (100% represents completing exactly on deadline), and presented as a histogram and cumulative distribution. Audio frames form a localized spike near 17% completion because there is little frame-to-frame variation and the overall workload is light. MPEG's distribution is more spread out and more heavy, while the GUI distribution is effectively bi-modal, with one low-computation clump and a few high-computation events (there is one GUI outlier that has been omitted, corresponding to the initialization sequence, which explains why the cumulative distribution is not shown reaching 100%).

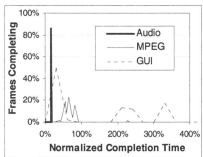 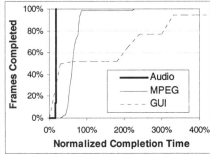

FIGURE 10.14 Frame Computation Histograms

A benchmark with a very consistent workload, such as Audio, will be more predictable, allowing the voltage scheduler to do a better job, while varied workloads, such as those of MPEG and GUI, make it more difficult for a system to anticipate future computation needs. Section 10.7 covers how these distributions interact with the dif-

Software and Operating System Support

ferent voltage scheduling algorithms, effecting the overall system energy consumed. A brief analysis of each application easily explains the observed distributions:

- **Audio** - The structure of Audio is very regular because the decryption method used, IDEA, is not data-dependant: a frame takes the same amount of time to compute independent of the input. If a data-dependent algorithm were used, or if the application was extended to include data-dependent compression, the frame distribution would broaden.

- **MPEG** - Internally, a MPEG video clip is composed of several different types of frames, each with a different function and processing requirement. Some frames, for example, represent a complete new scene, while others offer only incremental updates to a previous frame. Additionally, within each frame type there are variances depending on the exact composition of the scene.

- **GUI** - Different UI events generate drastically different workloads because they perform drastically different functions. For example, the work required to process a pen-down event is small, requiring only redrawing the surface of the button, while a pen-up event can be quite large because it may trigger a lengthy sequence of events, such as creating a new dialog box. The short events are much more common, but each long event contributes more to the overall energy consumed.

10.4.3 Task Failure

Benchmark success depends on the individual benchmark requirements. Audio and MPEG, similar data-flow applications, must sustain the required rate-of-flow, while the GUI application is concerned strictly with the quality-of-service presented to the user. A success metric is important for DVS studies because it sets a lower-bound on the processor speed; without it, it would be possible to run the processor at the slowest speed to realize maximum energy savings. Metrics such as ETR, attempt to factor out the 'minimum speed' requirement, which is useful for inter-architectural comparisons but does not work for software evaluation. The benchmark failure criteria method best represents real system behavior, but it does require the evaluation of specific benchmarks running on a complete system.

The Audio and MPEG benchmarks are required to maintain an aggregate frame-rate, a constraint that can be satisfied by not letting any single frame take too long to complete. Too many consecutive over-limit frames will overflow the output buffers, causing dropped frames and adversely effecting the quality-of-service. The system uses one output buffer per data-stream, allowing each frame to be tardy no more than one period. If a voltage scheduling algorithm slows the system such that the buffers overflow, then the algorithm fails and its energy result is not valid.

Failure for the GUI benchmark is more subtle, since it involves a quality-of-service judgement directly involving the user. If a hard all-frames-must-complete-by-X criteria is used then all scheduling algorithms would fail because there are some GUI frames with extremely long completion times. Therefore, we impose the rule that "all long-running GUI frames must execute at 40 MHz," which dictates that long-running frames will complete in a reasonable, standardized, amount of time without consuming excessive energy. This policy dictates a direct setting of 40 MHz for MFIXED, and for the other thread-based algorithms it imposes an over-deadline policy of 40 MHz (which is manually verified). The interval-based scheduling algorithms do not follow this policy, but their energy-consumption is always higher than MFIXED so we disregard this transgression. Speeds other then 40 MHz will significantly effect energy consumed by the GUI benchmark; however, it will effect all scheduling algorithms equally.

10.4.4 Workload Estimates

Thread-based voltage scheduling algorithms require a workload estimate, automatically formed by the operating system using a exponentially moving average defined by:

$$W_{NEW} = \frac{W_{OLD} \cdot k + W_{LAST}}{k+1}$$ (EQ 10.1)

Figure 10.15 shows the effect of varying the history weight, k. Increasing the exponentially moving average weight increases resistance to localized variance in the workload. This change can either increase or decrease the energy consumed, shown by Audio and MPEG, respectively.

A greater weight will produce estimates which mitigate local frame-to-frame variances, while smaller weights allow the system to adjust to short-term computational fluctuations. The optimal estimation weight depends on the system workload; a default weight of 6 is used as a compromise value. Theoretically, it is possible for individual applications to provide a frame-by-frame estimate of their workload based on application-specific information (for example, the *type* of MPEG frame about to be decoded); however, exploring this possibility was not undertaken.

When an new task first starts, the system will automatically run at full speed for a few frames, providing a statistical basis for workload prediction. Alternatively, an application can specify an initial workload estimate. Since the benchmark programs run for

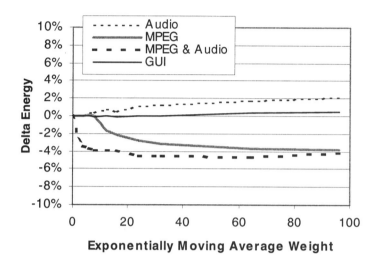

FIGURE 10.15 Workload Estimation Weight

many frames, eighty in the case of MPEG and over one hundred for Audio, this initial estimation period has little effect on overall behavior.

10.4.5 Frame-to-Frame Workloads

Frame-to-frame workload traces, shown in Figure 10.16, provide a detailed view of the benchmark characteristics. Both the frame distribution histogram and weighted workload estimates can be distilled from this view, which encompasses more raw data. Each graph consists of a primary trace, which is the un-averaged computation, and several derived traces, which are the exponentially-moving averages described in the proceeding section. Each sample point represents one frame of computation. For the Audio and MPEG benchmarks, therefore, the x-axis also represents time. For GUI, however, time is not constant; furthermore, the gap between frames is large enough to consider infinite.

These workload traces uncover another aspect of benchmark behavior: time-varying and cyclic patterns. A detailed analysis of the MPEG trace, for example, reveals that a roughly 3-frame cycle is present, with one high-computation frame followed by two lower-computation frames. Theoretically, this pattern could be used to generate better

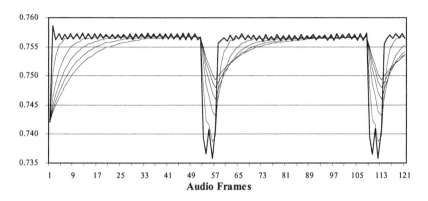

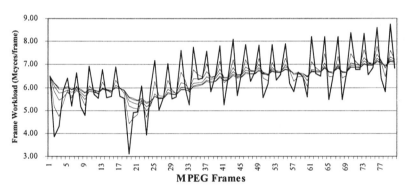

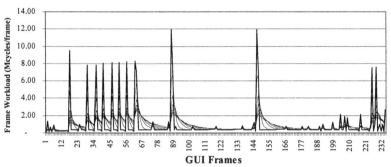

FIGURE 10.16 Benchmark Frame Workloads

Software and Operating System Support

workload estimates, resulting is lower energy consumption; exploiting this pattern, however, has not been undertaken in our work. The Audio benchmark, in contrast, is almost completely flat (the occasional dip corresponds to the end of the file), and the UI benchmark appears quite random: mostly low-cycles frames with a few large computations. Since workload is measured in terms of cycles, these characteristics are mostly independent of voltage scheduling and of any other tasks executing concurrently, they are artifacts of the underlying benchmark implementations (there will be small differences, however, due to cache and context-switch effects).

10.5 DVS Operating System

The DVS embedded operating system is based on the *armlib* run-time library (created and distributed by ARM Ltd.), which has been modified to support multi-threading and energy-efficient computing. Generically, an OS is responsible for managing shared global system resources, such as physical memory and processor cycles. The processor speed is one such resource, effecting the operation of all tasks in the system. Although an individual application may have a good idea of what speed it thinks it should run at, the operating system is ultimately responsible for determining the processor speeds which best serves *all* running tasks -- which is why the voltage scheduler is conceptually part of the operating system.

Only one process-level entity is supported by the run-time operating system; all global variables are system resources are shared and afforded equal access (programmable memory protection is not supported by the prototype MMU). Multi-threading support allows multiple virtual applications to execute within the monolithic process, yielding a variant of multiple processes which is suitable for our research. A basic Lock construct, similar to POSIX condition variables, is provided to synchronize communication between different threads.

The armlib OS is based on semi-hosted operation, where some functions, such as memory allocation, are performed locally while others, such as file operations, are handled with cooperation with a remote host over a communication link, in this case the direct serial debugging connection. This facility allows slow, but functional, access to a file system without requiring a physical disk; this technique is only useful for debugging since the serial connection will not be present during mobile operation. Data I/O is supported by a collection of system threads and libraries which process incoming data and distribute it to the appropriate data sink, providing the central clearing-house for the data supplied to the benchmark applications (raw MPEG or Audio data)

10.5.1 Energy-Efficient Functionality

Several custom operating system functions have been added to enable voltage scheduling, allowing a scheduler to determine and set the processor speed, as well as to control their own execution so they don't unnecessarily consume energy. The operating system calls implemented to control energy-related system aspects are summarized here:

Sys_Set_Speed Sets the system operating speed target, which is then communicated directly to the hardware. The actual operating speed, as well as the operating voltage, is determined dynamically by the hardware using a feedback loop which attempts to match the target operating speed. This function is never called by an application-level task, which specify their needs in terms of deadlines and execution requirements, and is only called directly by voltage schedulers.

Sys_Get_Counts Returns execution counts, such as number of active or sleep cycles, for the system as a whole since system startup. Internally, the system context-switch mechanism uses Sys_Get_Counts to attribute cycles to individual threads. A voltage scheduler then uses the per-thread counts, as well as the global counts, to determine how much work was performed by whom over a given time interval.

Sys_IdleWait Suspends the current thread until the system is about to enter into an idle state. The system will execute the suspended thread instead of entering the idle state; it is therefore necessary that a thread not dumbly loop on this call, else the idle state would never be entered. This function is primarily used by voltage schedulers when the system is executing at max speed to reduce the system overhead until there is a chance that the processor speed should be reduced, which will be indicated by a system idle interval.

Sys_ActiveWait The thread blocks and is signalled when the system leaves the idle state (the converse of Sys_IdleWait). This call can be used by a monitor or scheduling thread to go to sleep while the remainder of the system is idle and wait for some activity to monitor or schedule, preventing wasted energy.

A generic voltage scheduler, utilizing all these calls, is shown below. This outline is applicable to both interval-based and thread-based schedulers, with the

Calculate_Speed and Wait_Next_Iteration procedures changing depending on the algorithm used. Once a voltage scheduler has determined that the system is over-utilized (>= 100% target speed), or idle (<= 0% target speed), it can suspend the voltage scheduler until an appropriate change in system state. This is most important when the system enters a period of extended idle, when the scheduler should shut down (by calling Sys_ActiveWait) to prevent unnecessary scheduling (which would waste energy).

```
while(true)
{
    counts = Sys_Get_Counts()
    speed = Calculate_Speed(counts)
    Sys_Set_Speed(speed)
    if (speed >= 100%)
        Sys_IdleWait()
    if (speed <= 0%)
        Sys_ActiveWait()
    Wait_Next_Iteration()
}
```

10.5.2 Thread System

The thread system is responsible for multiplexing multiple execution contexts, called threads, onto the shared processor resource: it's primary job is to save and restore the processor state upon a context switch. In order to perform this job, it is necessary for the system to determine which thread should run when, a process called *temporal scheduling*. Additionally, the system must provide a mechanism for multiple threads to atomically access shared resources, which is accomplished through the use of a locking-semaphore construct. There are two main data structures used by the thread system, one for threads and one for locks. The thread data structure is primarily used to store context information when a thread is not active; the lock data structure is used to maintain information about who owns the lock as well as an ordered list of who is waiting for the resource.

Temporal scheduling is accomplished through a hybrid of two basic scheduling models commonly found in real-time systems *[10.8]*: priority-based and deadline-based. Conceptually, both models could be combined into one priority based scheme, where the priority of a deadline-based thread is determined by its relative deadline, but the implementation, and this discussion, treats the two models separately. New threads

DVS Operating System

default to a basic high-priority status, and may adjust themselves as necessary by supplying either a deadline or new priority.

Threads are fundamentally scheduled on a highest-priority first basis *[10.12]*, where priority is represented as a signed integer. Threads default to a priority of 0, and a typical high-priority is 100. Special system threads, such as the interrupt handler, execute at a higher priority (i.e. 105). For most system threads, the actual priority used is not important but the relative ordering is necessary to guarantee correct performance. Threads with identical priorities are scheduled on an Earliest Deadline First (EDF) basis. Threads without deadlines, or identical deadlines, are scheduled using a round-robin scheduler triggered by a preemptive interval timer maintained by the low-level OS.

The lock mechanism allows only one thread access to a specified piece of code at any given time. Any thread attempting to access the protected code while it is already locked is suspended and placed on an ordered list of threads waiting for that resource. When the current lock holder releases the resource, a waiting thread is activated and given access to the protected code. A similar mechanism allows a thread to wait for an event, suspending execution until it is triggered by another thread. The concept of temporal scheduling also applies to the ordering of threads waiting on locks: high priority threads are activated first, ensuring that they will run before their lower-priority counterparts.

The thread system runs independently of the voltage scheduler: it is concerned only with which thread should run next, not how fast it should run. However, it is very important that the voltage scheduler thread itself not be blocked for any great length of time. The worst case example is one of starvation: which can occur when the voltage scheduler determines that the processor should run at a slow speed due to lack of system activity and then the system becomes highly active due to a large amount of incoming network data. If the interrupt handler runs at a higher priority than the voltage scheduler, the system may become overloaded and fail before the voltage scheduler ever has a chance to execute and increase the processor speed. There are two software-only solutions to this problem. First, the voltage scheduler can always be run at the highest priority, ensuring that it will never be prevented from executing. Second, the interrupt handler or hardware can be used to increase the processor speed during an interrupt condition.

To prevent starvation of the voltage scheduler during periods of high interrupt activity, the hard target-speed register has two fields: normal operating speed and interrupt speed. When an interrupt condition is detected by the hardware, the system automati-

Software and Operating System Support

cally switches to the interrupt speed setting, typically programmed to the maximum processor speed. Therefore, although the voltage scheduler itself may not run during periods of high load, the processor will increase speed to maximum allowing the system to processing the incoming data without input buffers overflowing. Although this technique increases the interrupt processing bandwidth, it is not effective at decreasing the interrupt response time, as discussed in the next section. The hardware cost to provide this functionality is minimal, requiring only eight bits of register storage and a multiplexor to select between the two speed settings. The alternate to the hardware-based technique would be to embed a software-based speed increase in the interrupt handler, which also has a minimal cost but would spread the DVS implementation details throughout the operating system instead of keeping it constrained to the single voltage scheduler thread.

10.5.3 Interrupts and Timers

The ARM architecture defines two interrupts, FIQ and IRQ. The prototype implementation uses the FIQ interrupt for I/O transactions, reserving the IRQ interrupt for timer events and debugging operations. Interrupt handing is typically a tricky operation in embedded operating systems. The desire for a quick *interrupt response time* (IRT), the time taken from when the interrupt line is asserted until the task-level application handler begins execution, calls for a streamlined implementation that reduces overhead; the prototype system, however, implements a more general-purpose implementation and streamlines the programming interface. A single programmable timer interrupt is provided by the prototype hardware to enable time-based software events. This timer is managed by the operating system, which uses it for pre-emptive multitasking and to provide a sleep mechanism to application threads.

Interrupt handling in the prototype system is accomplished through a special FIQ_Wait function call, which atomically queues the calling thread and waits for a signal from the interrupt handler core. A external interrupt (call an FIQ in the ARM architecture) will trigger the interrupt handler, in turn signalling the waiting thread. Threads triggered by this mechanism execute as normal system threads and execute in their own memory space (which is not true of the interrupt handler routine itself); however, the two-level mechanism adds extra latency to the system. If an interrupt occurs and no threads are waiting, the interrupt is blocked until a handler thread becomes available and execution continues normally. The prototype hardware and software system was designed not to rely on a fast interrupt response, allowing for this added latency as well as the effects described below.

The Interrupt Response Time (IRT), is a critical quantity in many embedded systems. Furthermore, it is usually the worst-case IRT that is important in real-time systems, not the average-case. Unfortunately, DVS adversely affects the IRT because the worst-case behavior is determined by the slowest speed at which the processor can run. In the prototype system, IRT is bounded at about 100 cycles for the worst case, which is primarily determined by cache effects (multiple cache misses with writebacks, assuming a ½x external bus speed). The time of the response would therefore be on the order of 10 µs at 10 MHz, and 1 µs at 100 MHz, a factor of 10 difference. Additionally, there is the processing time required by the interrupt handler (which is application dependent). The worst-case low-to-high speed transition of the prototype system is approximately 70 µs, on the order of a context switch, which is quite long compared to the interrupt response time. Therefore, automatically changing the processor speed on an interrupt, as is done to prevent scheduling starvation, will not significantly effect the IRT. It is necessary to design the overall system to tolerate this wide variation in interrupt latencies; alternatively, one could limit the slowest allowable speed of the processor, which would increase energy consumption but reduce the IRT.

A hardware-supplied time-based interrupt mechanism is used by the operating system to process time-based events, such as pre-emptive multitasking and `Thread_Sleep` function calls. Since the hardware only supports one timer interrupt, software is responsible for aggregating all outstanding timer events and determining the next (soonest) trigger time. Threads waiting for a timer event wait for the hardware timer to trigger, when they are woken up and then continue processing. The pre-emptive multitasking timer functionality is used to maintain fairness in the presence of multiple tasks with identical priorities and deadlines. When this trigger fires, every 10 ms, the currently executing thread is suspended and the next thread begins execution.

10.5.4 I/O Processing Overhead

Some I/O and OS tasks required to support an application execute asynchronously of the application's frame processing; this work is not performed by the main application thread and is instead considered *overhead*. The overhead incurred, defined as all execution cycles not attributed to the main application thread, is shown in Figure 10.17 for each benchmark. Much of the overhead is caused by I/O data which is ultimately consumed by an application or generated by an application. Additionally, the OS must handle monitoring and scheduling multiple threads, which is necessary for functional operation but not directly part of any one application. Overhead is an important factor when evaluating voltage scheduling algorithms (how much

Software and Operating System Support

energy do we consume when scheduling the system); the overhead added by DVS scheduling is discussed in Section 10.7.

Data-flow through the prototype system, from the I/O interface to the eventual output device, is accurately modeled by the cycle-level simulator. There are several steps involved in the handling of the data, all of which are considered overhead except for the actual computation.

- **I/O Bridge** - Incoming external data is transferred to the I/O Bridge and places in an incoming-data FIFO, which stores data until the prototype CPU can read it and begin processing. The energy consumed by its internal architecture is not calculated, considered to be outside the processor subsystem, but the latency of the I/O Bridge activity is simulated. When data is present in the I/O Bridge device, it asserts a processor FIQ interrupt.

- **Interrupt Handling** - The prototype processor responds to an interrupt generated by the I/O device by activating the software interrupt handling routine, which reads I/O Bridge data and transfers it to an operating system owned data buffer. Data is packetized, based on information from the data transfer medium, and passed off to the data routing stage when a complete packet is formed.

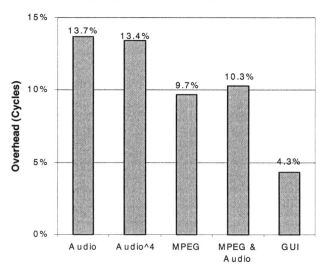

FIGURE 10.17 Benchmark Overhead

Energy Efficient Microprocessor Design

Work, measured as energy consumed, that cannot be directly attributed to the main application thread is considered overhead (measured here without any voltage scheduler running). Audio, for example, performs relatively little work decrypting each byte transferred, so the associated overhead is high; the per-byte computation for MPEG is higher so transfer overhead is lower; and the MPEG & Audio workload is dominated by the MPEG processing. The data transferred into the GUI system is very small, consisting of only a few bytes for a complete pen-event, resulting in a low overhead.

- **Data Routing** - The incoming data is routed to a type-based data buffer, where it held until the appropriate application can process it. Some data types are transferred directly to the application, when routing information is included in the packet header, while others require another layer of routing, such as a pen packet, which needs to be passed through the windowing system to determine which application receives the event based on the exact screen location of the event.
- **Computation** - Once data is placed in the application's incoming data buffer, it can be read and processed by the application; the output of the application, either in the form of graphical screen updates or raw audio data, is transferred to an output data buffer. The computation step comprises the bulk of execution cycles, and is the only one *not* considered overhead.
- **Data Output** - Data is held in the output buffer until it is time to be displayed. If data is output too early, the result is similar to over-deadline data: jerky video or choppy audio will result. The output buffer effectively reduces the jitter caused by the computation step to within limits tolerable by human perception and hardware output buffering. (This last step does not apply to the GUI application.)

10.6 Voltage Scheduling Algorithms

There are two forms of voltage scheduling algorithms, differentiated by the extent to which applications are required to inform the operating system of their behavior and requirements, evaluated here. *Interval-based* scheduling, which analyzes global processor utilization across fixed-size time intervals, requires no knowledge of individual application threads and attempts to adjust the system voltage based on aggregate system performance. *Thread-based* schemes, which perform somewhat better than interval based schedulers, require detailed knowledge of application behavior which must either be deduced by the operating system or provided by the application programmer.

The desired target operating speed, rather than a target operating voltage, is specified by a scheduling algorithm; the hardware is then responsible for setting the actual voltage to deliver the requested speed. The software is not aware of the actual operating voltage used by the hardware: the term "voltage scheduling" is used rather than "frequency scheduling" to remind us that the voltage is being adjusted along with the clock. In the following analysis a change in clock frequency implies an associated change in voltage, and vice-versa, unless otherwise noted. The speed to voltage translation is accomplished in hardware using a simple feedback mechanism: the current operating speed is measured against the target speed and then the current operating voltage adjusted either up or down, as appropriate.

The following sections describe the one off-line and four on-line scheduling algorithms used and give examples of their behavior.

10.6.1 Baseline Comparison (MFIXED)

MFIXED runs the processor at a pre-determined application-dependent fixed speed. For Audio and MPEG, this speed is empirically determined by running the processor at slower and slower speeds until the application fails; the GUI benchmark uses a fixed speed that provides a balance between performance and energy-efficiency. MFIXED is not a general-purpose run-time DVS algorithm because it can not dynamically adjust to varying workloads and requires a controlled environment and multiple executions to pre-determine the operating speed. A detailed description of the failure conditions for Audio and MPEG, used to determine the MFIXED speed, is discussed in Section 10.4.3. MFIXED is used as a comparison-point for evaluating the real dynamic voltage scaling algorithms: we would expect a "good" algorithm to perform comparably to MFIXED. Most on-line scheduling algorithms analyzed behave slightly worse than MFIXED, although in a few cases they perform better.

There are two fundamental problems with MFIXED, 1) it requires empirical pre-trial analysis to determine the fixed speed, and 2) it can not adapt to intra-benchmark workload variations. For example, the processing required to process an MPEG frame depend on the composition of the frame: complicated scene changes require more processing power. The MFIXED speed will be determined by the worst-case frame sequence, which could require considerably more processing than the average-case frame. However, the fixed-voltage representation does help us understand what should be possible with DVS; an omniscient on-line algorithm would be able to match MFIXED's performance.

10.6.2 Interval Based Scheduling (PAST, AVG)

The simplest dynamic scheduling approach, interval scheduling, attempts to determine the optimal voltage scheduling by analyzing global processor utilization across fixed-sized time intervals. Generally, if the processor is estimated to be highly loaded, the speed is increased, and if it is estimated to be lightly loaded, the processor speed is decreased. The major parameter governing the behavior of interval based scheduling is the duration of the measurement interval, which effects both the granularity of utilization measurements and the rate at which an algorithm will adapt to workload changes.

These algorithms are attractive in that they are very simple and non-intrusive to implement, requiring no code modifications at the application level. Their performance is very sensitive to the sampling interval used, which can be tuned to optimize for one specific application; however, the interval chosen may be inappropriate for another application, making it less useful for a general-purpose system executing multiple tasks. Additionally, the algorithms are often parameterized, i.e. with an averaging weight, which causes similar difficulties.

Interval-based scheduling algorithms make use of the *processor_utilization* variable, defined as the percent of time that the processor is active (not idle). This value is dependent on the processor speed: if the processor is 33% utilized while running at 100 MHz, it will be 66% utilized at 50 MHz, assuming a consistent workload.

Past Interval (PAST)

PAST adjusts the processor speed based on the processor utilization of the previous interval. If the utilization is above an upper threshold speed is increased; similarly for a lower threshold. PAST, which can be parameterized by its increment and threshold, reacts very quickly to changes in system workload: there is no memory of previous system utilization beyond the immediately preceding interval.

PAST Pseudocode:

```
each time_interval {
    if (processor_utilization > upper_threshold)
        newspeed = oldspeed + speed_adjustment;
    else if (processor_utilization < lower_threshold)
        newspeed = oldspeed - speed_adjustment;
}
```

Interval Average (AVG)

AVG uses a exponentially moving average to predict future utilization based on utilization history and is parameterized by *weight*, which dictates how quickly it adjusts to changes in utilization. Small values of weight, which can be as low as 0, emphasize a quick response time (making it behave similar to PAST), while large values make the system slow to respond to changes in processing requirements. The sub-expression "*oldspeed*processor_utilization*" calculates the instantaneous desired speed of the processor, which targets the system to run at 100% utilization (a 50% utilized 100 MHz system would slowly drift to 50 MHz).

AVG is fundamentally limited by its ability to estimate processor utilizations near 100%. For example, if the system is 50% loaded when running at 100 MHz, it will appear to be fully loaded when running *at anything less than 50 MHz*. Therefore, running at 20 MHz and 100% utilization is indistinguishable from running at 20 MHz and *over* 100% utilization: the algorithm can only approach the optimal speed from above. A functional implementation must always run at a higher speed than strictly necessary, wasting energy, so that it can retain the ability to detect situations when it should increase speed to handle a greater workload.

AVG Pseudocode:

```
each time_interval {
    if (processor_utilization >= 100%)
        newspeed = oldspeed + speed_adjustment;
    else
        newspeed = oldspeed*(weight + processor_utilization)/
(weight + 1);
}
```

10.6.3 Thread based scheduling (RATE, ZERO)

Thread-based voltage scheduling utilizes information about individual threads, provided in the form of real-time constraints, to calculate the target processor speed. Applications are divided up into chunks called *frames*, each of which has a measure of *work* which should be completed by it's *deadline*. A frame is an application-specific unit, such as a video frame, and work is defined in terms of processor cycles. The work required for each frame is automatically estimated by the system and updated to reflect the actual work performed as the application executes. The deadline, which indicates when the task should complete, and start-time, which indicates when it may begin, are specified by the application for each frame. The desired rate of processor execution in a single-threaded system is easily determined by the equation it is the job of the voltage scheduler to determine the processor speed given a set of multiple tasks

$$speed = \frac{work}{deadline - start} \qquad \text{(EQ 10.2)}$$

Two on-line algorithms have been simulated to explore our DVS system: RATE, adapted from *[10.15]*, and ZERO, which we have developed. Both algorithms reevaluate the processor speed each time a frame start-time or deadline is reached or a new thread is added to the system, conditions which could potentially change the processing estimate. The disadvantage of thread-based schedulers, compared to interval-based, is that they require application specified constraints and are more computationally intensive.

Average Rate (RATE)

The RATE algorithm assumes all tasks are periodic with a period derived from their start-time and absolute deadline, simply totaling the average rate required to sustain all tasks indefinitely

$$speed = \sum \frac{work_i}{deadline_i - start_i} \qquad \text{(EQ 10.3)}$$

RATE will optimally schedule a systems comprised purely of periodic tasks; however, sporadic tasks, which execute only once, and tasks with actual periods greater than their implied periods cause RATE to overestimate the processing required. This algorithm is computationally very simple and can be implemented in $O(n)$ time, where n is the number of scheduled threads.

TABLE 10.2 Sample Evaluation of the Rate Algorithm

	Start	Deadline	Period	Workload	Speed
Task A	0	3	3	120	40
Task B	2	4	3	30	15
Task C	4	6	NA	60	30

Table 10.2 shows the application of the RATE algorithm to a sample workload. Even though Task C is not periodic, the RATE algorithm reserves time in the schedule for it, causing the system to run faster than necessary, causing idle time. Furthermore, the processor speed is over-estimated since Task B's implied period, two, is greater than its actual period, which is three. Note that the result of the RATE algorithm does not depend on the relative performance of the tasks, i.e. if they are completing ahead or

behind schedule. So, for example, if the system is executing behind, such that all tasks are executing behind their deadlines, it makes no attempt to catch up. Given an accurate or over-estimated workload, the RATE algorithm will always complete tasks on or before deadline, if possible.

Zero-Start (ZERO)

The ZERO algorithm assumes all tasks are sporadic and calculates the minimum speed necessary assuming their relative start times are all zero. Given that threads are sorted in EDF order, this speed can be found by:

$$speed = \underset{\forall (i \leq n)}{MAX} \left| \frac{\sum_{j \leq i} work_j}{deadline_i - currenttime} \right| \qquad \text{(EQ 10.4)}$$

In general, ZERO tends to underestimate the amount of work required by the system, causing it to initially run slower than required, because it treats all tasks as sporadic: it does not consider the fact that a frame might be re-scheduled as soon as it finished. Conversely, the zero-start simplification causes ZERO to overestimate the processing required when a future task, one with a non-zero start-time, has a large amount of work to be completed. Overall, the underestimation tends to be greater than the overestimation and as will be see will actually be found to be beneficial. The algorithm can be implemented in *O(n)* time where *n* is the number of scheduled threads; the required list of sorted threads can be maintained incrementally as task deadlines are updated. Without the zero-start assumption, the computational complexity would increase significantly, requiring a complex on-line scheduling algorithm *[10.15]* which would not significantly improve performance

Table 10.3 shows the operation of the ZERO algorithm. In contrast to RATE, ZERO ignores the implied task periods, causing it to under-estimate the processing speed required. In this example, the system initially calculates the speed to be 40, just enough to complete Task A by its deadline. Since Task A is periodic, it is re-inserted into the schedule when it completes. At this point, reevaluating the ZERO algorithm indicates that the system should run at a speed of 68. Had ZERO taken into account the re-application of Task A, it could had initially run at a more efficient speed, somewhere between 40 and 68.

TABLE 10.3 ZERO Algorithm with t=0 and t=3

t=0	Start	Deadline	Period	Workload	Sum(Work)	Speed
Task A	0	3	3	120	120	40
Task B	2	4	3	24	144	36
Task C	4	6	NA	60	204	34
t=3						
Task B	(-1)	1	3	24	24	24
Task C	1	3	NA	60	84	28
Task A	0	3	3	120	204	68

10.6.4 Simulation Traces

Figure 10.18 shows a trace of each of the scheduling algorithms running on the beginning of the MPEG benchmark. From this, one can start to see the how the different scheduling algorithms behave in practice. Only the trace for the primary MPEG thread is shown, all other tasks, those considered overhead, are excluded. Occasionally, some algorithms show processing speed spikes reaching 100 MHz; these spikes represent the return from another thread, typically the interrupt handler, which has set the processor speed to maximum. Vertical traces within a block of computation indicate a brief context switch away from the main thread of computation

The ZERO Algorithm assumes all tasks are sporadic and uses an approximation algorithm to determine the immediate processor speed, which is the maximum of the speed calculated for each task group (a task and the tasks before it).

The MAX trace, representing the max speed of 100 MHz, highlights the separate MPEG frames running at 125 ms/frame; however, when the processor slows down it is hard to delineate individual frames. The PAST and AVG algorithms are shown with a sampling interval of 140 ms; AVG slowly settles on a speed which is slightly too fast, but PAST over-estimates the processing required and starts to run at fullspeed. RATE and ZERO both settle on a medium speed, with RATE slightly overestimating (shown by periods of idle time) and ZERO initially underestimating and then ramping up speed to make up for lost cycles. ZERO is the most effective algorithm on this workload, discussed more in Section 10.7.

Software and Operating System Support

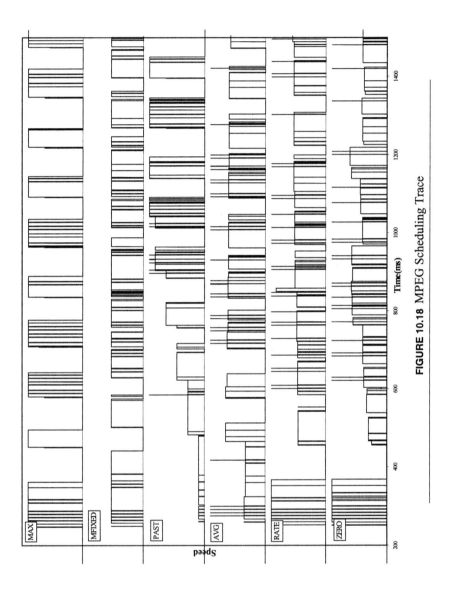

FIGURE 10.18 MPEG Scheduling Trace

336   *Energy Efficient Microprocessor Design*

10.6.5 Target-Deadline Scheduling

In a traditional hard real-time system, it is desirable to complete all tasks before their deadline. However, in a voltage scaling system it is wasteful to complete a task before its deadline if the system would otherwise be idle: it would be better to run the system at a slower speed and complete the task exactly on deadline. Unfortunately, tasks are not fully predictable and exhibit frame-to-frame execution variance; therefore, requiring them to complete before their deadline would entail scheduling using their worst-case execution time, resulting in higher energy consumption. *Target real-time* treats deadlines as intermediate checkpoints instead of strict requirements, requiring the system to tolerate missed deadlines but minimizing the energy consumed. Target real-time is different than semantics traditionally found in real-time systems: *hard real-time*, which means that it is absolutely necessary to complete by deadline, and *soft real-time*, which means that it would be nice to complete by deadline, but it is not absolutely necessary.

A task's required output behavior can still achieved while consuming the minimum amount of energy by dividing them into two parts: the core that represents the bulk of the work would be executed using target real-time, while the output buffer stage, discussed in Section 10.5, can be executed using soft real-time. This way, the majority of execution cycles can be scheduled more efficiently, reducing overall energy consumption, while the output-buffering stage can be scheduled more aggressively, providing the required output semantics. The difficulty of voltage scheduling in hard real-time systems is explored in *[10.10]*.

Theoretically, executing using target real-time is a trade-off between delay and energy: task delay is increased by adding additional buffering to decrease the overall energy consumed. The input- and output- buffering mechanisms result in three data frames being present in the system at any time (for the multi-media tasks): one waiting in the input-buffer, one being processed by the application, and one waiting for output. The amount of buffering required depends on the characteristics of the application. An application with very high frame-to-frame variance might benefit from increased buffering, while others might not require any buffering at all. The exploration of this concept, which is not undertaken, could potentially yield significant benefits.

10.6.6 Scheduling of Non-Deadline Tasks

The primary goal of the thread-based voltage scheduler is to schedule deadline-based threads; however, there are several situations where threads without deadlines will exist and need to be factored into the scheduling. Some tasks, such as interrupt handlers or event dispatchers, consist of a small amount of work that does not have a deadline but should be executed before other threads because they may in turn activate a thread that does have a deadline. Other tasks may involve batch processing, representing a large unpredictable amount of work, that need to be completed eventually but not necessarily anytime soon. These categories by no means provide coverage across all possible task types; however, they are sufficient for the applications running on the software system described.

High-priority tasks are executed by the temporal scheduler before other non-high-priority tasks, and are executed in highest-priority order. High-priority tasks, however, are ignored by the voltage scheduler, which assumes that they are likely to complete before the system could change speed. The Extra Work extension, described below, attempts to predict the future requirements of high-priority tasks. The other possible behavior would be for the system to increase execution speed, most likely to maximum, when a high-priority task is executing; except, if all high-priority tasks are assumed to be short-lived then this is not necessary. A system containing longer-running latency-critical high-priority tasks would have to modify its scheduling policy, but the implemented mechanism is adequate for the DVS system.

Rate-based threads are mapped onto the deadline-based scheduler by assigning a relative deadline which is automatically updated whenever the task executed. Tasks specify a desired rate of execution, in terms of cycles/second; the initial deadline is supplied by the operating system. Each scheduling interval, a new deadline is calculated using $deadline_{new} = deadline_{old} + work\_accomplished/desired\_rate$, where $work\_accomplished$ is the number of cycles the thread executed in the proceeding interval. This mechanism guarantees a sustained rate of execution and avoids starvation (except if caused by a high-priority task).

10.7 Algorithm Analysis

This section presents the results of the voltage scheduling algorithms running in the simulation environment. Primarily, inter-algorithm comparisons are made, showing how the different algorithms behave on a particular benchmark. In some cases, the

Algorithm Analysis

parameterization of a particular algorithm is studied, to understand how that parameter effects operation. Furthermore, extensions designed to increase energy efficiency of the thread-based scheduling algorithms are described and analyzed. Results are normalized to the processor running at full speed and compared against the baseline algorithm MFIXED to contrast DVS with static-voltage operation.

10.7.1 Evaluation

The evaluation benchmarks are executed using a cycle-level simulator that is aware of the current system operating speed and voltage. The complete software system necessary to support the benchmarks (voltage scheduler, complete operating system, and low-level interrupt handlers) are included in the simulation to accurately measure their effect. Executing the voltage scheduling algorithms adds additional scheduling overhead that consumes additional energy, which is minor compared to the gains.

The voltage scheduling algorithms are evaluated using the total energy consumed for a given execution configuration normalized to the total energy consumed by the processor running at full speed, allowing us to compare the effectiveness of the different algorithms as well as to measure the overall savings afforded by DVS. Furthermore, results are subject to the constraint that all applications execute correctly, so if a particular configuration results in task failure that result is not shown in a graph.

Figure 10.19 shows the energy consumed and processor speed used by MFIXED for each fundamental benchmark and a few selected combinations, normalized to the

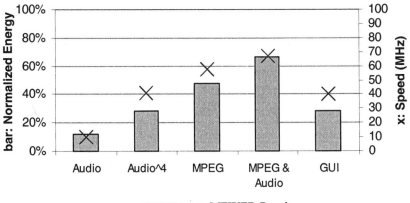

FIGURE 10.19 MFIXED Results

energy consumed by the processor running at max-speed. The differences in these energy numbers represents fundamental characteristics of the benchmark applications, not the effects of a *dynamic* run-time scheduling algorithm: the Audio benchmark presents a much lighter workload than MPEG, allowing a reducing the operating voltage and energy/operation. There is a loose correlation between the processor speed and energy consumed, because some system effects, such as overhead and cache behavior, are non-linear. Since the results are normalized to the processor running at full speed, the effect due to different numbers of instructions executed for the various benchmarks is normalized out.

From the MFIXED results it can be seen that the gains achievable by DVS are extremely application dependent. Some, such as Audio, show a great potential for low-energy operation; others, which require full processing power, do now allow their speed to be reduced, and therefore will show limited reduction in energy consumed. However, for all cases, DVS reduces the total energy consumed by the system by a significant amount.

10.7.2 Voltage Scheduling Results

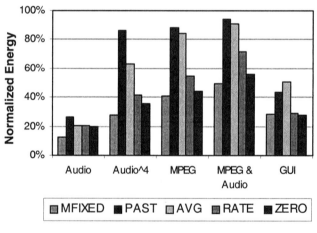

FIGURE 10.20 Composite Results

Figure 10.20 summarizes the energy consumed while running the presented scheduling algorithms. MFIXED consumes the least energy for all benchmarks except GUI, for which ZERO consumes slightly less; RATE and ZERO do fairly well when com-

Algorithm Analysis

pared with MFIXED; PAST and AVG perform poorly overall. The interval scheduling algorithms, PAST and AVG, have difficulty adapting to the different benchmarks: AVG, for example, performs comparably with RATE for Audio, but does a horrible job at adapting to MPEG.

The energy consumed by each benchmark running at the minimum feasible speed, normalized to the energy consumed by the processor running at full speed, indicates the potential saving afforded by DVS. "Audio^4" indicates that four instances of the Audio benchmark are being run simultaneously, while "MPEG & Audio" indicate one instance of MPEG and one of Audio.

Figure 10.21 shows the effect of the ZERO algorithm on the frame completion time for the three benchmarks. The weighted distributions are shown, so the area under the curve represents the total time taken by the benchmark to complete. The frame completion time for a given frame is increased when the processor speed is reduced, moving the distributions to the right. The other voltage scaling algorithms, apart from ZERO, are omitted for clarity (their behavior is similar).

The normalized energy of increased workloads increases because of the associated increase in operating speed: a 100% utilized system would be at 100% normalized energy. Since the energy consumed in each case is normalized to the energy of the system running at max-speed, the increase in energy due to an increased number of execution cycles is normalized out. DVS provides the best gains for lightly loaded system; these gains disappear as the system becomes increasingly utilized.

10.7.3 Multiple Benchmark Execution

Executing multiple benchmarks at the same time, shown in Figure 10.22, explores the ramifications of voltage scheduling within a multi-tasking environments. Because the maximum speed of the processor is limited, it is not possible to run two instances of the MPEG decoder at the same time. Additionally, the GUI benchmark is incompatible with Audio and MPEG because of its large fraction of idle time: the energy consumed by either Audio or MPEG will dominate that of GUI, making the comparison uninteresting. In general, the voltage scheduler behaves as expected, adapting the processor speed to accommodate the extra workload; however, there are several non-obvious side-effects, discussed below.

Figure 10.23 shows how increasing the processor workload actually *decreases* the number of missed deadlines. This effect is counter-intuitive: generally, in fixed-speed systems, increasing the processor workload is disadvantageous because it results in

Software and Operating System Support

Software and Operating System Support

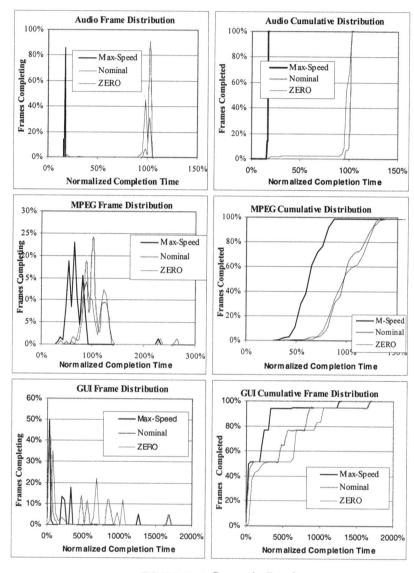

FIGURE 10.21 Composite Results

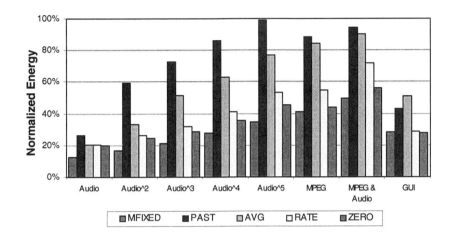

FIGURE 10.22 Multiple Benchmark Execution

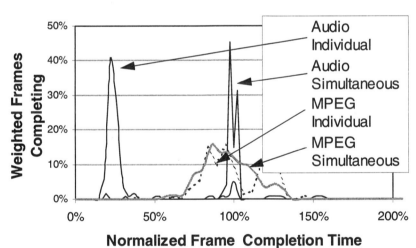

FIGURE 10.23 Effect of Processor Loading on Frame Completion

more missed deadlines in a static-speed system. The underlying reason for the decreased missed deadlines in the DVS system is the increase in processor speed cause by multiple benchmark execution: from the point of view of Audio, the processor is running faster so its frames complete quicker. The benefit is less realized for MPEG because its execution is delayed by the execution of the Audio frames (which tend to have earlier deadlines). This phenomenon is another demonstration of the fundamental energy/latency trade-off: arbitrarily increasing the processor speed would also reduce the number of over-deadline frames at the expense of increased energy.

The work of the voltage scheduling algorithm and other operating system structures also increases with an increased workload. Figure 10.24 shows the overhead for sev-

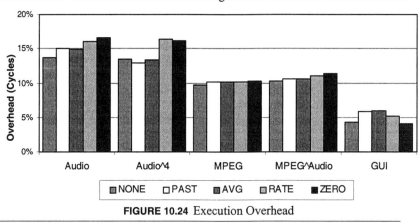

FIGURE 10.24 Execution Overhead

eral benchmark combinations. It is interesting to note that the system overhead is *reduced* by applying a voltage scheduler in some cases: slowing the processor speed down can cause multiple small data-transfer blocks to be lumped together into one larger block, reducing overhead. The overhead of the interval-based algorithms (PAST, AVG) is actually reduced by an increased workload: higher processor speeds mean more non-scheduling instructions per interval, effecting a reduction in the overhead ratio. The thread-based algorithms are effected by the amount of work per frame, and therefore perform more work when multiple benchmarks execute.

10.7.4 Interval Scheduling Parameterization

The interval parameterization of the PAST and AVG algorithm has a great effect on their performance, shown in Figure 10.25. At a fundamental level, the weakness of interval-based scheduling algorithms revolves around the requirement to have this system-level parameter: an interval which performs well for one application will not necessarily perform well for another.

Algorithm Analysis

If the interval is too short an interval-based algorithm will increase speed too quickly and not find the average case speed, wasting energy; if the interval is too long the algorithm will be slow in adapting to changing requirements, either wasting energy or causing the application to fail (as is the case for the MPEG & Audio benchmark combination). There is no one "correct" interval for a system capable of executing several different applications; therefore, any fixed interval will be sub-optimal for a generic workload. The parameterization of *upper_threshold*, *lower_threshold*, *speed_adjustment*, and *weight* incur problems similar to the interval parameterization and are not analyzed in detail.

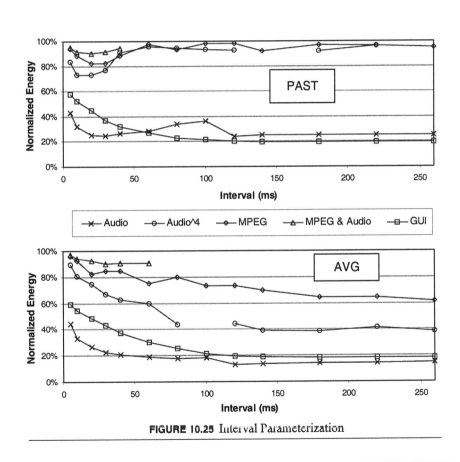

FIGURE 10.25 Interval Parameterization

10.7.5 Algorithm Extensions

The thread-based scheduling algorithms have been extended in several ways to improve energy efficiency:

- **Schedule Smoothing** - Because of our basic voltage scheduling assumptions, frames often complete at or near their deadline; however, the value computed by the simplistic scheduling model $rate = work_{remain}/(deadline - current\_time)$ becomes very large as the absolute deadline is approached (the denominator approaches zero). Instead, schedule smoothing causes threads to be scheduled such that the current frame *and* the next frame should complete by the *following* deadline, that is $rate = (work_{remain} + work_{next})/(deadline + period - current\_time)$. Schedule smoothing has the effect of smoothing out the frame-to-frame variance while still dynamically adjusting to the workload, significantly reducing energy (Figure 10.26).

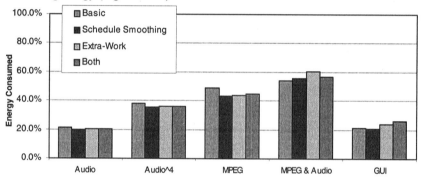

FIGURE 10.26 Effects of Schedule Smoothing and Extra-Work Prediction

In the case of MPEG, which has considerable frame-to-frame variance, it can reduce energy by as much as 12%. Neither Audio nor GUI, which have relatively stable distributions, have their energy greatly decreased (but they are not adversely effected, either). The MPEG & Audio benchmark shows a slight increase in energy due to Schedule Smoothing (3.3%). All other results incorporate Schedule Smoothing.

- **Extra Work Prediction** - As discussed earlier, execution cycles attributed to high-priority threads are by default ignored by the voltage scheduler. The Extra Work option counts the cycles not attributed to scheduled threads and factors them into the scheduled rate calculation, predicting future non-accounted cycles based on past behavior. The slight increase in processor speed should then cause tasks to complete closer to their stated deadline when they would have otherwise been delayed by spurious high-priority task execution.

Extra Work does have the effect of decreasing missed task deadlines, at the expense of increased energy (Figure 10.26); however, its utility is questionable because the applications have been architected to tolerate missed deadlines and increasing the processor speed increases energy consumption. In general, it is desirable to have the system run slightly behind, in fact to always miss the deadline.

- **Over-Deadline Policy** - Deadline-based real-time systems must deal with the question of what to do when a task exceeds its deadline. With hard real-time, system failure occurs; for soft real-time it is sometimes desirable to terminate the current frame and begin execution of the next, and sometimes desirable to ignore the situation and continue. Under target real-time, however, it is desirable to continue processing; furthermore, with a DVS system it is necessary to determine the speed of execution.

  The default over-deadline policy supplied by the simple scheduling model is to continue executing at maximum processor speed (otherwise, the denominator would become non-positive, and the result would be negative or undefined). It is possible, however, to specify that when a task exceeds its deadline it should be switched over to a rate-based model with an application-supplied rate parameter. The over-deadline policy is not necessarily mutually exclusive with schedule smoothing, as the over-deadline speed would be invoked when the task exceeded twice it's deadline (however, since our periodic tasks fail when they exceed twice their deadline this situation does not occur). The over-deadline policy is designed mainly for applications similar to GUI, which have a few long-running frames without hard failure characteristics; without the extension, the GUI benchmark has greatly increased energy consumption.

- **Event Filtering** - Formulating an averaged-case execution workload estimate for predicting future frame execution assumes that the event workloads follow a roughly normal distribution with reasonable variance. However, the GUI benchmark distribution has several outliers that artificially increase the workload estimate used; furthermore, these outliers are extremely large and could never complete by their deadline even when the processor is running at full speed. Therefore, the Event Filtering option restricts the workload estimate process to only utilize frames with feasible completion times. Event Filtering only signifi-

Software and Operating System Support

cantly effects the GUI benchmark, seen in Figure 10.27, as the other benchmarks either contain no (Audio) or only one (MPEG) outlier. All results are with event-filtering enabled by default.

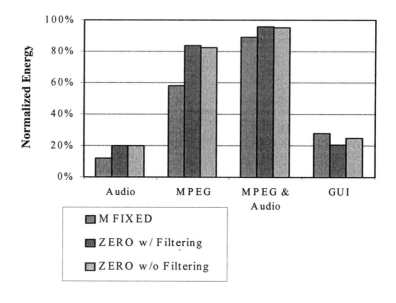

FIGURE 10.27 Effects of Event Filtering

10.8 Comments and Possible Further Directions

The prototype system provides a DVS solution which performs well for the implemented application set. However, there are several areas which have been uncovered in the actual operation of the DVS system. First, extending the thread-based model to more complicated systems with intermixed applications will introduce dependencies that will not be handled correctly by our implementation. Second, a complete implementation of the system has shed new light upon the assumptions on which it was based which did not include task chaining.

Comments and Possible Further Directions

Our implementation of thread-based scheduling is prone to application chaining: Task A triggers the execution of Task B, while the important deadline is the completion of Task B. In our current system, the future execution requirements of the second task, which is marked as non-runnable while it waits for a signal, will not be factored into the speed calculation. What completion deadline should Task A specify? It cannot specify the same deadline as Task B, else Task B might not start until it is too late. Furthermore, the system can't always schedule Task B as runnable, because it does not know that Task A will trigger Task B. Therefore, the system requires some mechanism for specifying a task chain that would allow the scheduler to perform the necessary computation. The software environment did not contain any such task chaining, so it is not a solution space that has been explored.

A real-time task typically assumes that the operating system will attempt scheduling so as to complete tasks before their deadline. As a practical matter, however, tasks in the DVS system are allowed to complete after their deadline (relying on extra output buffering to complete correctly). Furthermore, it can be seen that it is more efficient for a periodic task to *always complete after its deadline*, avoiding energy-wasting idle-time (if the system were idle, it could have been run slower). It is unclear from a cursory analysis exactly how far after the deadline a task should complete for optimal performance, a question whose answer requires the capture and analysis of the statistical distributions involved. Furthermore, altering the fundamental task deadline-completion assumption potentially alters many basic real-time operating system theorems (such as "is EDF scheduling optimal"), requiring them to be re-visited in the context of a DVS system.

References

[10.1] T. Truman, T. Pering, R. Brodersen, "The InfoPad multimedia terminal: a portable device for wireless information access," *IEEE Transactions on Computers*, vol. 47, 1998.

[10.2] M. Weiser, "Some computer science issues in ubiquitous computing," *Communications of the ACM*, Vol. 36, pp. 74-83, July 1993.

[10.3] M. Weiser, B. Welch, A. Demers, and S. Shenker, "Scheduling for reduced CPU energy," *Proc. 1st Symp. on Operating Systems Design and Implementation*, pp. 13-23, Nov. 1994.

[10.4] *ARM 8 Data-Sheet*, Document Number ARM DDI0080C, Advanced RISC Machines Ltd, July 1996

[10.5] T. Burd and R. W. Brodersen, "Energy efficient CMOS microprocessor design," *Proc. 28th Hawaii Int'l Conf. on System Sciences,* Vol.1, pp. 288-297, Jan. 1995

[10.6] SSLeay cryptography library, obtained from <http://www.psy.uq.oz.au/~ftp/Crypto>.

[10.7] Y. Endo, Z. Wang, J. B. Chen, and M. Seltzer, "Using Latency to Evaluate Interactive System Performance," *Proc. 2nd Symp. on Operating Systems Design and Implementation*, Nov. 1996.

[10.8] A. Burns and A. Wellings, *Real-Time Systems and Programming Languages, second edition,* Addison-Wesley, 1997.

[10.9] gd gif-manipulating library, version 1.2, obtained from <http://sunsite.unc.edu/boutell/index.html>.

[10.10] I. Hong, M. Potkonjak, M. Srivastava. On-Line Scheduling of Hard Real-Time Tasks on variable Voltage Processor. IEEE/ACM International Conference on Computer-Aided Design, Nov 1998.

[10.11] C. J. Linbald and D. L. Tennenhouse, "The VuSystem: A Programming System for Compute-Intensive Multimedia," IEEE Journal of Selected Areas in Communication, 1996.

[10.12] C. Liu and J. Layland, "*Scheduling algorithms for multiprogramming in a hard real-time environment*", CACM 20, 1973.

[10.13] B. Shneiderman, *Designing the User Interface*, Addison-Wesley, 1992.

[10.14] Tcl version 1.5, obtained from <http://www.sunlabs.com/research/tcl>.

[10.15] F. Yao, A. Demers, and S. Shenker. A scheduling model for reduced CPU energy. In IEEE Annual Foundations of Computer Science, pages 374-382, 1995.

CHAPTER 11 *Conclusions*

11.1 Energy Efficient Design

Processor systems are widely prevalent in portable devices which demand increasingly higher levels of energy-efficiency. However, processor energy-efficiency has lagged behind custom ASIC's and DSP chips, such that while the processor carries only a fraction of the computational load, it is a significant, if not dominant, component of the overall system energy consumption. In order to address this problem it is necessary to use a design methodology that incorporates energy consumption as a primary consideration in all stages of the design process, from the software down to the circuit design. A critical component of this methodology is maximizing the degrees of design freedom by taking into account the characteristics of the application. If this is done, it is possible to significantly improve processor energy-efficiency, thereby enabling smaller, more powerful, and longer running portable devices.

The first step is to develop a set of metrics for energy-use efficiency, and benchmarks which capture the requirements of the application domain under consideration. The energy-throughput ratio, ETR, captures not only the energy efficiency but the performance (throughput) realized by the processor. This was enhanced to the BETR metric in order to properly include the effects of energy consumption during the periods of time when performance is not required at all. The bursty nature of the computation, which is a common characteristic to many portable devices, could best be exploited

Conclusions

by the use of Dynamic Voltage Scaling (DVS) which was shown to allow an order of magnitude improvement in the ETR.

While DVS can be very effective, it is only one component of the solution to energy efficiency. While the optimal solution might be to start from the beginning and design a new instruction set, the practical requirements of software compatibility required that a standard ISA be used and we chose the instruction set from Advanced Risc Machines, Ltd. Given the ISA the next step is to analyze the basic hardware architecture of the processor and to determine the organization which best optimizes the ETR metric. It was found that enhancements which used sophisticated instruction-level parallelism rarely improved the performance enough to compensate for the increased energy required for their implementation. Optimization of the memory sub-system was found to be critical, as well as low-energy techniques for dealing with the interchip communications.

While circuit-level optimizations do not provide the magnitude of improvement which can be achieved by architectural and system optimizations, they remain important. In particular, the design of circuit blocks which minimize switching, as well as operate at low voltages and function robustly against dynamic variations of these voltages requires new design considerations. An energy-conscious design flow is therefore required which enables energy consumption optimization at all levels, from the high-level behavioral simulation to the final verification strategies which incorporate the requirements of dynamic, low voltage operation.

Flexible supply voltage generation is a critical component of energy efficiency and it is therefore necessary to provide flexible DC-DC conversion with high efficiency, particularly at low voltage. To fully exploit DVS, it is imperative to provide this conversion across a broad range of supply voltage, as well as to change between voltage levels with switching transients on the order of microseconds, while incurring minimal energy consumption as overhead. CMOS switching regulators, with sophisticated digital control, can provide the solution to this need.

DVS also requires modification of the software, but the critical modifications can be implemented via a modular voltage scheduler, and easily integrated into the operating system. Interval-based scheduling, which analyzes global processor utilization across fixed-size time intervals, requires no additional modification to the operating system or application software. However, utilizing thread-based scheduling, which performs better than interval-based scheduling, does require knowledge of application behavior, which must either be deduced by the operating system or provided by the application programmer, and therefore requires further changes to the software infrastructure in order to fully realize the potential of DVS.

11.2 Current Industry Directions

In the rapidly evolving processor industry, some of the techniques described in this book are beginning to come to fruition. Of particular interest is run-time voltage/frequency adaptation, which was not even considered feasible three or four years ago within the industry, and yet, is rapidly emerging in a variety of products technologies:

In 1999, Intel introduced SpeedStep, which runs the processor at two different voltages and frequencies, depending upon whether the notebook computer was plugged into an AC outlet, or running of its internal battery.

In 2000, Transmeta introduced LongRun, which dynamically varies voltage and frequency over the range of 1.2-1.6V and 500-700MHz, providing a 1.8x variation in processor energy consumption. Control of the voltage/frequency is in firmware, which monitors the amount of time the operating system is sleeping.

In 2000, AMD introduced PowerNow!, which dynamically varies voltage and frequency over the range of 1.4-1.8V and 200-500MHz, providing a 1.7x variation in processor energy consumption. Control of the voltage/frequency is implemented via a software driver which monitors the operating system's measure of CPU utilization.

In 2001, Intel introduced the XScale processor, which is essentially the second generation StrongArm. It can dynamically operate over the voltage and frequency range of 0.7-1.75V and 150-800MHz, providing a 6.3x variation in processor energy consumption, the most aggressive range announced to date. By further advancing the energy-efficiency of the original StrongArm, this device will be able to deliver 1000 MIPS with an average power dissipation as low as 50mW at 0.7V, yielding an effective MIPS/Watt as high as 20,000.

While many energy-efficient circuit design techniques, as well as some architectural design techniques, have been realized in commercial products in recent years, the potential of optimizing a microprocessor from its inception, without any legacy constraints, has not yet been achieved. Utilizing an energy-conscious, top-down design flow to fully optimize the architecture, processor system, and even software may yet yield further dramatic improvement in the energy efficiency of commercially-available microprocessors.

11.3 Future Directions

Integrating the voltage converter onto the same chip as the processor is a straightforward enhancement of the system, which would not only reduce cost, but would also enable further optimizations. This integration has already been demonstrated on other research devices. Further optimizations would include the ability to have multiple supply voltages in the system with little additional cost.

One potential use of an additional supply would be for the external processor bus, which could then operate at a speed independent of the processor core. This would allow high-speed DMA to the main memory, so that even when the processor core is operating at low speed, high-bandwidth I/O-memory transactions could still occur. Additional areas would be applying DVS to external I/O devices, which could likewise dynamically trade-off performance (bandwidth) versus energy consumption.

Another direction would be the further exploration of instruction set architecture and microarchitecture for improving energy-efficiency. Of particular interest are VLIW and parallel processor architectures which explicitly expose their parallelism, and do not suffer from exponentially increasing energy consumption with parallelism as is the case with superpipelined and superscalar processor architectures.

As process technology continues to advance, energy consumed by interconnect will consume an increasingly larger fraction of the total energy consumption. Thus, further investigation of low-swing interconnects could yield additional improvement of processor system energy-efficiency.

The software support for DVS is an area requiring additional research and improvement. Extending the thread-based model to more complicated systems with intermixed applications will introduce dependencies that are not handled correctly by our implementation. In addition, systems which run applications without hard, real-time constraints (e.g. notebook computers and PDAs) can relax the constraints imposed by the application's real-time deadlines, which may alter basic real-time operating system theorems, requiring them to be re-visited in the context of a DVS system.

Index

A
ALU 81
ARM
    ARM V4 ISA 126
    ARM710 30, 32
    ARM8 49, 65, 72, 73, 76, 158, 171, 175, 302
    ARM8 architecture 313
    ARM810 76, 303
    ARM940T 76, 303
    pipeline 176

B
Benchmark
    AUDIO 300, 311
    Dhrystone 297
    GUI 311
    MPEG 300, 311
    UI 300
Buck Converter 218
Burst-mode ETR (BETR) 27-9
Bus interface 198

C
Cache 58, 59, 60, 192-3
    associativity 65
    busses 187
    characteristics 186
    controller 173, 190
    double Reads 73
    level-0 68
    policies 66
    replacement policy 67
    sequential reads 73
    tag memory 62
    write buffer 71, 174, 195
    write miss policy 67
CAM 64, 96
Chip photos
    DVS processor 170
    interface IC 285
    regulator 276
    SRAM 212
Clock Control Domains 173
Clocks
    drivers 87, 139
    skew 144
    wiring 144
Conversion efficiency 245
Coprocessor 201
CPL 80

D
Data Flow 172
Datapath 90
    cell 155
DMA 174
Domino 80
DSP 3, 79
    architectures 129
DVS regulator chip 157
DVS system 307
Dynamic logic 117
Dynamic Voltage Scaling (DVS) 4, 34, 54, 125 292

E
Efficiency 245, 273
E_{IDLE} 27
E_{MAX} 23
Energy Breakdown 185
Energy breakdown 184
Energy estimation 131
Energy per Operation 19
Energy-Delay Product (EDP) 23

Index

Energy-to-Throughput (ETR) 23, 24, 26, 31, 33

F
FIQ 326
frequency detector 261

G
Gating 85
Ground Bounce 210

I
I/O interface 48, 328
Idle energy 297
InfoPad 306
Instruction set architecture (ISA) 49
Interface chip 285
Interrupts 326
IRQ 326

L
Latches 87, 137
Layout optimizations 90
Loop filter 264
Loss mechanisms 274

M
Memory 92, 211
Metrics
    SPECint92 30, 32
    SPECint95 21
MIORFT 84
MIPS R4700 31, 32

N
Noise margin 112, 150
NORA 80
Notebook computers 57

O
Operating system 322

P
PathMill 165
PDA 9, 26, 57

Palm-PC 3
Phase Locked Loop (PLL) 58, 114
Power Bounce 210
PowerMill 128, 161
PowerPC 603 57

R
RAM 47
Register file 99
Registers
    Load/Store 74
Regulator Interface 205
Routing
    global 210
    local supply 154
    power distribution 151
    signal 156
    supply distribution 151

S
Scheduling
    interval 331, 344
    thread 332
Scheduling algorithms
    MFIXED 330
    PAST 331
    Rate 333
    smoothing 346
    ZERO 334
Short-crcuit current 83
Signal processing 55
Structural simulation 160
Sleep mode 57, 75
SRAM 61, 93, 120, 212
Standard cell layout 91
StrongArm 2, 176, 290, 302-3
Superpipelined 53
Superscalar 51, 55
Supply bounce 113
Switching Regulators 217
    boost 238

buck 218
buck-boost 240
dissipation 223, 231
PFM 227
PWM 219
Synchronous rectifier 279
System architecture 284

T

Task failure 318
T_{AVE} 27
Test vector generation 158
Thread system 324
TimeMill 161
Timing
    analysis 129, 165
    race condition 142, 147, 149
    verification 161
    violations 150
T_{MAX} 27
Transceiver
    bus 100
    interchip 104
    intrachip 107
    low-swing 103
Tri-state busses 119

V

VCO 206, 207
Velocity saturation 20
Verification 129
    behavioral 157
    path Identification 163
    test vector generation 158
    timing 161
VHDL 129
    structural 159
Voltage scheduler 292, 308, 325

W

Workload 319